CAMBRIDGE LIBRARY COLLECTION

Books of enduring scholarly value

Botany and Horticulture

Until the nineteenth century, the investigation of natural phenomena, plants and animals was considered either the preserve of elite scholars or a pastime for the leisured upper classes. As increasing academic rigour and systematisation was brought to the study of 'natural history', its subdisciplines were adopted into university curricula, and learned societies (such as the Royal Horticultural Society, founded in 1804) were established to support research in these areas. A related development was strong enthusiasm for exotic garden plants, which resulted in plant collecting expeditions to every corner of the globe, sometimes with tragic consequences. This series includes accounts of some of those expeditions, detailed reference works on the flora of different regions, and practical advice for amateur and professional gardeners.

Catalogues of Plants in the Dublin Society's Botanic Garden, at Glasnevin

This plant catalogue is in two parts. The first, published in 1801, provides a list, organised according to the Linnaean system, of the hothouse and greenhouse plants in the newly established Botanic Gardens at Glasnevin, near Dublin, and the second, from 1802, is a guide to all the plants in the gardens, including the arboretum and 'Hortus tinctoria', where dye plants (important for the Irish linen industry) were displayed, arranged according to the layout of the different beds, and giving their Linnaean and common names. Both are believed to have been written by Walter Wade (1740–1825), a physician and enthusiastic botanist who had lobbied for the establishment of a botanical garden on the site of the former property of Thomas Tickell, poet and government administrator in Ireland. The gardens were inaugurated in 1795, and these catalogues form the earliest evidence of their considerable holdings of native and exotic plants.

Cambridge University Press has long been a pioneer in the reissuing of out-of-print titles from its own backlist, producing digital reprints of books that are still sought after by scholars and students but could not be reprinted economically using traditional technology. The Cambridge Library Collection extends this activity to a wider range of books which are still of importance to researchers and professionals, either for the source material they contain, or as landmarks in the history of their academic discipline.

Drawing from the world-renowned collections in the Cambridge University Library and other partner libraries, and guided by the advice of experts in each subject area, Cambridge University Press is using state-of-the-art scanning machines in its own Printing House to capture the content of each book selected for inclusion. The files are processed to give a consistently clear, crisp image, and the books finished to the high quality standard for which the Press is recognised around the world. The latest print-on-demand technology ensures that the books will remain available indefinitely, and that orders for single or multiple copies can quickly be supplied.

The Cambridge Library Collection brings back to life books of enduring scholarly value (including out-of-copyright works originally issued by other publishers) across a wide range of disciplines in the humanities and social sciences and in science and technology.

Catalogues of Plants in the Dublin Society's Botanic Garden, at Glasnevin

WALTER WADE

CAMBRIDGE
UNIVERSITY PRESS

CAMBRIDGE
UNIVERSITY PRESS

University Printing House, Cambridge, CB2 8BS, United Kingdom

Cambridge University Press is part of the University of Cambridge.
It furthers the University's mission by disseminating knowledge in the pursuit of
education, learning and research at the highest international levels of excellence.

www.cambridge.org
Information on this title: www.cambridge.org/9781108081535

This edition first published 1801
This digitally printed version 2017

ISBN 978-1-108-08153-5 Paperback

The original edition of this book contains a number of oversize plates
which it has not been possible to reproduce to scale in this edition.
They can be found online at www.cambridge.org/9781108065580

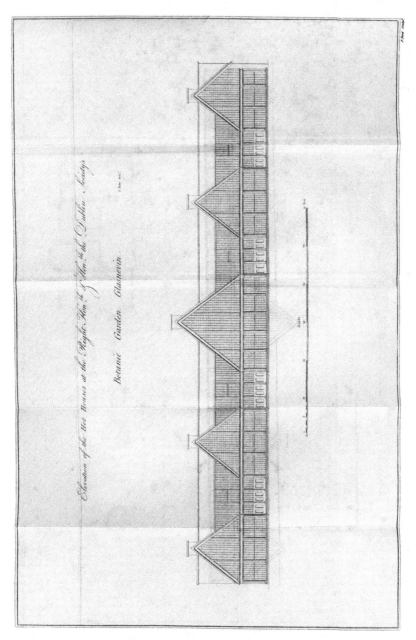

Elevation of the Hot Houses at the Right Hon.ble & Plan to the Dublin Society

Betani.c Garden Glasnevin.

The material originally positioned here is too large for reproduction in this reissue. A PDF can be downloaded from the web address given on page iv of this book, by clicking on 'Resources Available'.

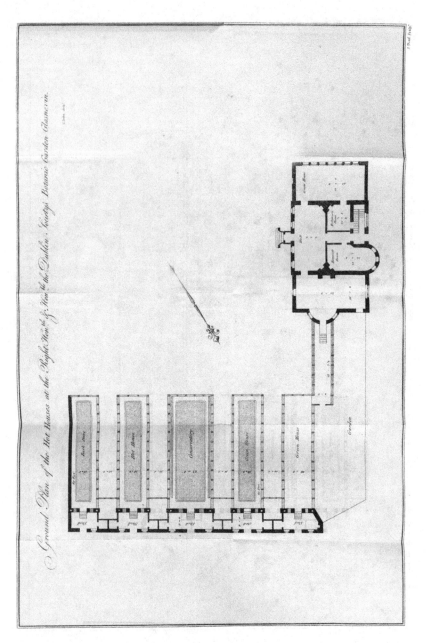

The material originally positioned here is too large for reproduction in this reissue. A PDF can be downloaded from the web address given on page iv of this book, by clicking on 'Resources Available'.

SYSTEMATIC

CATALOGUE

OF

Green and Hothouse Plants,

IN

THE DUBLIN SOCIETY'S

BOTANICAL GARDEN

AT

GLASNEVIN;

WITH

THEIR SCIENTIFIC NAME, COMMON NAME, COUNTRY OR
NATIVE SOIL, TIME OF FLOWERING, AND WHETHER
ANNUAL, BIENNIAL, PERRENNIAL, TREE OR SHRUB.

𝔇𝔲𝔟𝔩𝔦𝔫:

PRINTED BY GRAISBERRY & CAMPBELL, 10, BACK-LANE

———

1801.

ABBREVIATIONS EXPLAINED.

S.	Stove, or hot houfe.
G.	Green houfe.
C. G. H.	Cape of Good Hope.
M. P. Y.	Moft part of the year.
M. P. S.	Moft part of the fummer.
V. S.	Various feafons.
an.	Annual.
bi.	Biennial.
pe.	Perennial.
tr.	Tree or fhrub.

CLASSIS I.

MONANDRIA MONOGYNIA.

Scientific Name.	Common Name.	Country, or Native Soil.	Time of Flowering. &c.	
			M. P. S.	*S. pe.*
CANNA	INDIAN SHOT			
1. Indica.	common	Asia, Africa, America,	*M. P. S.*	*S. pe.*
v. striata.	Striated do.			
v. lutea.	Yellow do.			
2. glauca.	glaucous leaved.	S. America.	July.	*S. pe.*
3. flaccida.	flaccid.	Carolina.		*S. pe.*
RENEALMIA	RENEALMIA			
1. nutans.	nodding.	E. India.	March.	*S. pe.*
AMOMUM	GINGER			
1. Zingiber.	narrow-leaved.	E. India, naturalized to Jamaica.	September.	*S. pe.*
2. Zerumbet.	broad-leaved.	Indies.		*pe.*
3. Granum paradis.	Grains of Paradise.	Madagascar, Guinea, Zeylon.		*S. pe.*
COSTUS	COSTUS			
1. speciosa.	superb Costus.	E. India.		*S. pe.*
MARANTA	ARROWROOT			
1. arundinacea.	Indian.	S. America.	August.	*S. pe.*

MONANDRIA MONOGYNIA.

CURCUMA	TURMERICK			
1. longa.	long rooted. -	E. India,	August.	S. pe.
KAEMPFERIA	KAEMPFERIA			
1. Galanga.	Galangle. -	Indies.	June, July.	S. pe.
THALIA	THALIA			
1. geniculata.	jointed. ▪	S. America.		S. pe.
LOPEZIA	LOPEZIA			
1. Mexicana.	Mexican. ▪	Mexico.	October.	S. an.
POLLICHIA	POLLICHIA			
1. campestris.	whorl leaved. ▪	C. G. H.	September.	G bi.

CLASSIS II.

DIANDRIA MONOGYNIA.

JASMINUM	JASMINE			
1. Sambac.	Arabian.	Indies.	M. P. S.	S. tr.
v. fl. plen.	Double flowered ditto.			tr.
v. Tuscanic.	Tuscany ditto. -			
2. vimineum.	twiggy. -	Java, Malabria.		S. tr.

DIANDRIA MONOGYNIA.

JASMINUM	JASMINE			
3. glaucum.	glaucous.	C. G. H.	Aug.	G. tr.
4. Azoricum.	Azorian.	Madeira.	M. P. S.	G. tr.
5. odoratissimum.	yellow Indian.	Madeira.	May, October.	C. tr.
6. grandiflorum.	Catalonian.	Indies.	June, Oct.	S. tr.
OLEA	OLIVE			
1. Europea.	European.	S. of Europe.	July.	G. tr.
var. latifol.	broad leaved ditto			G. tr.
var. buxifol.	box leaved ditto.		July.	G. tr.
2. Capensis.	Cape.	C. G. H.		G. tr.
var. undulata.	wave leaved ditto			G. tr.
3. fragrans.	sweet scented.		July.	G. tr.
4. Americana.	American.	Cochinchina, China, Japan.	July.	G. tr.
		Carolina and Florida.	June.	G. tr.
FONTANESIA	FONTANESIA			
1. phillyreoides.	phillyrea leaved.	Syria.		tr.
JUSTICIA	JUSTICIA			
1. Ecbolium.	long spiked.	Arabia, Malabar, Ceylon.	May, August.	S. tr.
2. coccinea.	scarlet flowered	Cajenne, South America.	March.	S. tr.
3. nasuta.	dichotomous.	India.	M. P. S.	S. tr.
4. pectoralis.	forked.	Domingo, Martinico.		S. p.
5. Adhatoda.	Malabar nut.	Ceylon.	June.	G. tr.
6. Peruviana.	Peruvian.	Lima.	M. P. S.	S. tr.

U

DIANDRIA MONOGYNIA

JUSTICIA	JUSTICIA.			
7. hyssopifolia.	Snap tree.	Canary Islands.	M. P. S.	G. tr.
8. orchioides.	broom leaved.	C. G. H.	August.	G. tr.
9. parviflora.	small flowered.	Calcutta.		G. pe.
CALCEOLARIA	CALCEOLARIA.			
1. pinnata.	winged leaved.	Peru.	M. P. S.	S. an.
2. Fothergillii.	spatula leaved.	Falkland Islands.	June.	G. bi.
VERBENA	VERVAIN.			
1. nodiflora.	creeping.	Nepolis Cicily, East Indies, Carabee Islands, Virginia.	M. P. S.	S. pe.
2. triphylla.	three leaved.	Chili, Bonaria.	August.	G. tr.
3. Aubletia.	cut leaved rose.	Virginia.	M. P. S.	G. bi.
SALVIA	SAGE			
1. bicolor.	two coloured.	Barbary.	June.	G. pe.
2. Canariensis.	Canary.	Canary Islands.	Oct.	G. tr.
3. Africana.	African.	C. G. H.	April, June.	G. tr.
4. aurea.	gold flower'd African.	Ditto.	May, Nov.	G. tr.
5. paniculata.	panicled.	Africa.	June August.	G. bi.
6. pinnata.	winged leaved.	Crete.		G. tr.
7. Mexicana.	Mexican.	Mexico.	May, July.	G. tr.
8. rugosa.	wrinkle leaved.	C. G. H.		

DIANDRIA MONOGYNIA.

SALVIA	SAGE.			
9. violacea.	violet flowered.	- West India.	April, May.	*S. tr.*
10. formosa.	shining leaved.	- Peru.	*M. P. S.*	*G. tr.*
11. coccinea.	scarlet flowered.	- East Florida.		*S. tr.*
ERANTHEMUM	ERANTHEMUM.			
1. pulchellum.	blue flowered.	- East India.	January, May.	*S. tr.*

CLASSIS III.

TRIANDRIA MONOGYNIA.

OXYBAPHUS	OXYBAPHUS			
1. viscosus.	clammy.	- Peru.	June, September.	*S. pe.*
IXIA	IXIA.			
1. bulbifera.	bulb bearing.	- C. G. H.	April.	*G. pe.*
2. purpurea.	purple.	-	July.	*G. pe.*
3. rubeo-cyana.	swans neck.	-	May.	*G. pe.*
4. reflexa.	reflex flowered.	-		*G. pe.*
5. capitata.	headed.	-	April.	*G. pe.*
6. crispa.	curled.	-	June.	*G. pe.*

U 2

TRIANDRIA MONOGYNIA.

IXIA.

IXIA.					
7. spicata.	spiked.	C. G. H.	—	May.	G. pr.
8. cinnamomea.	cinnamon.	—	—	June.	G. pr.
9 fistulosa	hollow leaved.	—	—	April.	G. pr.
10. tricolor.	three coloured.	—	—	May.	G. pr.
11. erecta.	upright.	—	—	June.	G. pr.
12. polystachia.	many spiked.	—	—	July.	G. pr.
13. spectabilis.	shewy.	—	—	May.	G. pr.
14. deusta.	copper coloured.	—	—	—	G. pr.
15. crocata.	crocus flowered.	—	—	April.	G. pr.
16. aristata.	bearded.	—	—	June.	G. pr.
17. conica.	conic.	—	—	May.	G. pr.
18. longiflora.	long flowered.	—	—	—	G. pr.
19. maculata.	spotted.	—	—	May, June.	G. pr.
20. propinqua,		—	—	—	G. pr.
21. flexuosa.	bowed.	—	—	—	G. pr.
22. incarnata.	flesh coloured.	Carolina.	—	—	G. pr.
23. coelestina.	beautiful.	C. G. H.	—	—	G. pr.
24. corymbosa.	corymbus.	—	—	—	G. pr.
25. rosea.	rose.	—	—	—	G. pr.

TRIANDRIA MONOGYNIA.

GLADIOLUS		C. G. H.		G. p.
1. triftis.	CORN FLAG. fquare ftalked.	.	May, June.	G. p.
2. longiflorus.	long flowered.	.	June, July.	G. p.
3. galeatus.	helmet flowered.	.		G. p.
4. rofeus.	rofe.	.	May, June.	G. p.
5. verficolor.	various coloured.		June.	G. p.
6. ringens.	gaping.	.	May,	G. p.
7. precox.	early.		April.	G. p.
8. polyftachius.	many fpiked.	.	May, July.	G. p.
9. gramineus.	grafs leaved.	.	May, Auguft.	G. p.
10. Merianellus.			May, June.	G. p.
11. Merianus.				G. p.
12. anguftatus.	narrow leaved.	.		G. p.
13. fecuriger.	copper coloured.			G. p.
14. tubiflorus.	tube flowered.	.		G. p.
15. blandus.	blufh coloured.	.		G. p.
16. plicatus.	plaited.	.		G. p.
17. ftrictus.	upright blue.	.		G. p.
18. pulchellus.				G. p.
19. fuperbus.	fuperb.	.	May.	G. p.
20. cardinalis.	large flowering.		May, July.	G. p.

TRIANDRIA MONOGYNIA.

ANTHOLYZA					
1. ringens.	gaping.	-	C. G. H.	May, June.	G. pe.
2. Cunonia.	scarlet.	-	——	——	G. pe.
3. Æthiopia.	broad leaved.	-	——	——	G. pe.
4. plicata.	plaited.	-	——	April.	G. pe.
5. spicata.	spiked.	-	——	June.	G. pe.

N. B. Many beautiful *varieties*, besides the *species* mentioned of Ixia, Gladiolus and Antholyza.

ARISTEA					
1. cyanea.	grafs leaved.	-	C. G. H.	June, July.	G. pe.
IRIS, or *Flower de Luce.*					
1. tricufpis.	three pointed.	-	C. G. H.		G. pe.
2. fisyrinchium.	crocus rooted.	-	Spain, Portugal, Barbary.		G. pe.
3. Martinicenfis.	Martinico.	-	Martinico.	June, Auguft.	G. pe.
4. pavonia.	peacock.	-	C. G. H.		G. pe.
5. Chinenfis.	Chinefe.	-	China.	May, June.	G. pe.
6. bulbifera.	bulb bearing.	-			
7. longifolia.	long leaved.	-	C. G. H.	May, June.	G. pe.
8. maculata.	fpotted ditto.	-			
MORÆA.					
1. cærulea.	blue.	‹	C. G. H.		G. pe.

TRIANDRIA MONOGYNIA.

MORÆA.				
2. iriopetala.	Iris flowered.	- C. G. H.	June.	G. pe.
3. iridioides.	fword leaved.	- Conftantinople.	—	G. pe.
4. Northiana.	broad leaved.	- S. America.	—	G. pe.
CYPERUS.				
1. alternifolius.	alternate leaved.	- Madagafcar.	March.	S. pe.
2. denudatus.	naked.	- C. G. H.		G. pe.
WACHENDORFIA				
1. thyrfiflora.	thyrfe flowered.	- C. G H.	May, June.	G. pe.
KYLLINGIA.				
1. triceps.	three headed.	- America, and E. India.	M. P. S.	S. pe.
2. umbellata.	umbelled.	- Eaft India.	July, Auguft.	S. pe.
TAMARINDUS				
1. Indica.	Tamarind tree.	- Egypt, E. & W. Indies.	June, July.	S. pe.

DIGYNIA.

SACCHARUM.			
1. officinarum.	SUGAR CANE. common,	- Indies.	S. pe.
PANICUM	PANIC GRASS.		
1. polygamum.	Guinea-grafs.	- W. India.	S. pe.

TRIANDRIA DIGYNIA.

DOG'S TAIL GRASS.

CYNOSURUS				
1. Ægyptius.	Egyptian.	Africa, Asia, America.	July, September.	S. an.
2. Indicus.	Indian.	Indies.		S. an.

CLASSIS IV.

TETRANDRIA MONOGYNIA.

PROTEA.

PROTEA.				
1. decumbens.	decumbent.	C. G. H.		G. tr.
2. conocarpa.	tooth leaved.	———		G. tr.
3. purpurea.	purple.	———		G. tr.
4. Scolymus.	small smooth leaved.	———	May, June.	G. tr.
5. mellifera.	honey bearing red backed.	———	Jan. Feb.	G. tr.
6. pallens.	pale.	———	June.	G. tr.
7. conifera.	cone bearing.	———	April, June.	G. tr.
8. sericea.	silky.	———		G. tr.
9. argentea.	silvery.	———		G. tr.
10. grandiflora.	great flowered.	———	August.	G. tr.
11. speciosa.	shewy, or oblique leaved.	———	June, July.	G. tr.

TETRANDRIA MONOGYNIA.

PROTEA.				
11. speciosa.	shewy.	•		June, July.	**G. tr.**
var. alb.	white leaved.				
var. nigr.	black ditto.				
12. hirta.	hairy.		C. G. H.		G. tr.
13. spathulata.	flat leaved.			July, August.	G. tr.
14. umbellata.	umbelled.				G. tr.
15. torta.	twisted leaved.			July, August.	G. tr.
16. parviflora.	small flowered.			August.	G. tr.
17. tomentosa.	woolly.				G. tr.
18. plumosa.	feathered.			June, August.	G. tr.
19. pulcella.	fennel leaved.		New Holland.	June, July.	G. tr.
20. incurva.	incurved.		C. G. H.		G. tr.
21. glabra.	smooth.			June, July.	G. tr.
22. racemosa.					
23. Levisanus.	branching.	•		April, July.	G. tr.
24. corymbosa.					
25. linearis.	linear leaved.	•			G. tr.
26. pubera.					
27. alba.	white.	•			G. tr.
BANKSIA.					
1. ferrata.	saw leaved.		New Holland.		G. tr.
2. dentata.	tooth leaved.	•			G. tr.

X

TETRANDRIA MONOGYNIA.

BANKSIA	**BANKSIA**			
3. ericæfolia.	heath leaved.	-	New Holland.	May, August.	G. tr.
4. præmorsa.	snipped.	-	——		G. tr.
5. spinulofa.	spiny.	-	——		G. tr.
PERSOONIA	**PERSOONIA**			
1. lucida.	shining.	-	New Holland.	July, August.	G. tr.
2. linearis.	linear leaved.	-	——	——	G. tr.
EMBOTHRIUM	**EMBOTHRIUM**			
1. buxifolium	box leaved.	-	New Holland.		G. tr.
2. sericeum.	silky.	-	——	M. P. Y.	G. tr.
GLOBULARIA	**GLOBULARIA**			
1. longifolia.	long leaved.	-	Madeira.	July, August.	G. tr.
SCABIOSA	**SCABIOSA**			
1. rigida.	Æthiopian, or rough leaved	Æthiopia.		G. tr.
2. Africana.	African.	-	Africa.	July, October.	G. tr.
3. Cretica.	Cretan.	-	Crete.	June, October.	G. tr.
OPERCULARIA	**OPERCULARIA**			
1. aspera.	rough.	-	New Zealand.	June, July.	G. pe.
HOUSTONIA	**HOUSTONIA**			
1. coccinea.	scarlet flowered.	-	W. Coast of America.	June, August.	G. tr.
SIPHONANTHUS	**SIPHONANTHUS**			
1. Indica.	Indian.	-	S. America.		S. tr.

TETRANDRIA MONOGYNIA.

CATESBÆA	LILY-THORN			
1. fpinofa.	fpiny.	Ifle of Providence, and others of the Bahama.	M. P. S.	S. tr.
IXORA.	IXORA			
1. coccinea.	fcarlet.	E. India.	June, Auguft.	S. tr.
2. alba.	white.			S. tr.
PAVETTA	PAVETTA			
1. Indica.	Indian.	E. India.	Auguft, October,	S. tr.
BUDDLEA	BUDDLEA			
1. falvifolia.	fage leaved.	C. G. H.	Auguft, Sept.	G. tr.
EXACUM	EXACUM			
1. vifcofum.	clammy.	Canary Iflands.	June, July.	G. b.
PLANTAGO	PLANTAIN			
1. Capenfis.	Cape.	C. G. H.	M. P. S.	G. tr.
MONETIA.	MONETIA.			
1. Barlericides.	four fpined.	E. India, C. G. H.	July.	S. tr.
DORSTENIA	DORSTENIA			
1. Contrajerva.	angular leaved.	Mexico, Peru, Tobago, Ifland St. Vincent.	M. P. S.	S. p.
CURTISIA	CURTISIA			
1. faginea.	Haffagay tree.	C. G. H.		G. tr.

X 2

TETRANDRIA MONOGYNIA.

CHLORANTHUS				
1. inconfpicuus.	tea leaved, or *Chu-lan.*	China, Japan.	*M. P. S.*	*G. tr.*
CAMPHOROSMA				
1. Monfpeliaca.	Montpelier.	Spain, Narbone, Tartary.	Auguſt, Sept.	*G. tr.*
RIVINA				
1. humilis.	downy.	Caribee Iſlands, Jamaica.	*M. P. S.*	*S. tr.*
2. laevis.	fmooth.	Barbadoes, America.	———	*S. tr.*
LAMBERTIA				
1. formofa.	red flowered.	New Holland.	July.	*G. tr.*
2. longifolia.	long leaved red flowered.	Botany Bay.	———	*G. tr.*
CONCHIUM				
1. nervofum.	nerved.	New Holland.	June, July.	*G. tr.*
2. faligmum.	willow leaved.	———	———	*G. tr.*
3. gibbofum.	downy.	———	———	*G. tr.*
4. aciculare.	needle leaved.	———	———	*G. sr.*
5. latifolium.	broad leaved.	———	———	*G. tr.*

TETRAGYNIA.

MYGINDA			
1. uragoga.	pubefcent.	S. America	*S. tr.*
2. latifolia.	broad leaved.	———	*E. tr.*

CLASSIS. V.

PENTANDRIA MONOGYNIA.

HELIOTROPIUM	TURNSOLE		
1. Peruvianum.	Peruvian.	•	Peru.	*M. P. Y.*	*S. tr.*
2. Indicum.	Indian.	•	W. India.	July, Auguſt.	*S. an.*
3. parviflorum.	ſmall flowered.	•	W. India.		*S. an.*
BORAGO	BORAGE				
1. Indica.	Indian.	•	E. India.	June, Oĉtober,	*S. an.*
ECHIUM	VIPER'S BUGLOSS				
1. fruticoſum.	ſhrubby.	•	Æthiopia.	May, June.	*G. tr.*
2. candicans.	hoary tree.	•	Madeira.	May.	*G. tr.*
3. giganteum.	gigantic.	•	Teneriff.	May, July.	*G. tr.*
4. ſtriĉtum.	upright.	•	————	*M. P. S.*	*G. tr.*
TOURNEFORTIA	TOURNEFORTIA				
1. volubilis.	climbing.	•	Jamaica, Mexico.	July, Auguſt.	*S. tr.*
2. hirſutiſſima.	very hairy.	•	S. America.		*S. tr.*
CYCLAMEN	CYCLAMEN				
1. Perſicum.	Perſian.	•	Iſland of Cyprus.	Feb. April.	*G. p.*
MENYANTHES	BUCK BEAN				
1. ovata,	oval leaved.	•	C. G. H.	May, June.	*G. p.*

PENTANDRIA MONOGYNIA.

ANAGALLIS				
1. Monelli.	Italian.	Italy.	July, Sept.	G. pe.
STYPHELIA				
1. tubiflora.	tube flowered.	New Holland.		G. tr.
PLUMBAGO	LEAD-WORT			
1. Zeylanica.	Ceylon.	E. India.	April, Sept.	S. tr.
2. rosea.	rose coloured.		June, August.	S. tr.
CONVOLVULUS	BIND-WEED			
1. Scammonia.	Scammony.	Spria, Mysia, Cappadocia.	May, August.	S. pe.
2. Canariensis.	Canary.	Canary Islands.	June, August.	G. tr.
3. althaeoides.	mallow leaved.	Levant, Sicily.	M. P. S.	G. tr.
4. linearis.	linear leaved.			G. tr.
5. Cneorum.	silvery leaved.	Spain, Crete, Syria.	May, August.	G. tr.
6. spithamaeus.		Virginia.		S. an.
7. faxatilis.	rock.	Spain, Italy.		G. pe.
8. panduratus.	Virginian.	Virginia.		G. pe.
9. Hermanniae.	Hermannia's.	Peru.		G. pe.
10. pentaphyllus.	five leaved.	W. India.	August, Sept.	S. an.
11. Turpethum.	square stalked.	Ceylon		S. pe.
IPOMÆA				
1. Quamoclit.	winged leaved.	E. India.	July, August.	S. an.
2. tamnifolia.	tamnus leaved.	Carolina.		S. an.

PENTANDRIA MONOGYNIA.

CAMPANULA.	**BELL-FLOWER.**				
1. mollis.	soft.	-	Syria, Sicily, Spain.	May, August.	G. pe.
2. aurea.	golden.	-	Madeira.	August.	G. tr.
3. versicolor.	various coloured.	-			G. pe.
4. rupestris.	rock.				G. pe.
TRACHELIUM.	**THROAT-WORT**				
1. cæruleum.	blue.	-	Italy, Levant.	July, Sept.	G. bi.
RONDELETIA	**RONDELETIA**				
1. Americana.	American.	-	America.		S. tr.
SOLANDRA	**SOLANDRA**				
1. grandiflora.	great flowered.	-	Jamaica.	March, April.	S. tr.
GOODENIA	**GOODENIA**				
1. ovata.	oval.	-	New Holland.	June, Oct.	G. tr.
2. laevigata.	smooth.	-	——	——	G. tr.
3. calendulacea.	marygold.	-	——	August.	G. tr.
CINCHONA	CINCHONA or *Jesuits Bark-Tree.*				
1. Caribaea.	Caribean.	-	W. India.		S. tr.
COFFEA	**COFFEE TREE**				
1. Arabica.	Arabian.	-	Arabia Felix, Æthiopia.	August, Oct.	S. tr.
CHIOCOCCA	**SNOW-BERRY**				
1. racemosa.	opposite-leaved.	-	Jamaica, Barbadoes.	September.	S. tr.

PENTANDRIA MONOGYNIA.

HAMELLIA	HAMELLIA			
1. patens.	spreading.	Hispaniola.	July, Auguſt.	S. tr.
2. grandiflora, vel ventri-coſa.	great flowered.	Jamaica.	September.	S. tr.
MIRABILIS	MARVEL of PERU			
1. dichotoma.	forked.	Mexico.	July, Auguſt.	S. pe.
DATURA	THORN APPLE			
1. Metel.	hairy.	Aſia, Africa, Canary Iſlands.	July, Oct.	S. an.
2. laevis.	ſmooth capſuled.	Abyſſinia.	July,	S. an.
3. arborea.	tree.	Peru.	Mar. July, Aug.	S. tr.
HYOCYAMUS	HENBANE			
1. aureus.	ſhrubby.	Crete, Levant.	March, Oct.	G. tr.
NICOTIANA	TOBACCO			
1. fruticoſa.	ſhrubby.	C. G. H., China.	July, Auguſt.	S. tr.
ATROPA				
1. fruteſcens.	ſhrubby.	Spain.	March.	G. tr.
PHYSALIS	WINTER-CHERRY			
1. tomentoſa.	downy.	C. G. H.	July, Auguſt.	G. tr.
2. Barbadenſis.	Barbadoes.	Barbadoes.	April, Oct.	S. an.
3. Peruviana.	Peruvian.	Lima.		S. tr.
4. arboreſcens.	arboreſcent.	Campechia.		S. tr.
5. proſtrata.	trailing.	Peru.	Auguſt	S. an.
6. ſomnifera.	cluſtered.	Spain, Mexico,	July, Auguſt.	G. tr.

PENTANDRIA MONOGYNIA.

SOLANUM	**NIGHTSHADE**			
verbafcifolium.	mullien leaved.	America.	June, Sept.	G. tr.
1. Pfeudo-Capficum.	winter cherry.	Madeira.	June, July.	G. tr.
2. Melongena.	egg plant.	Afia, Africa, America.	June, July.	S. an.
3. marginatum.	white	Abyffinia.	June, Auguft.	G. tr.
4. tomentofum.	woolly.	Æthiopia.	June, July.	G. tr.
5. laciniatum.	cut leaved.	New Zealand.	July, Auguft.	G. tr.
6. fodomeum.	black fpined.	Africa.	June, July.	G. tr.
7. Indicum.	Indian.	Indies.	July.	S. tr.
8. finnuatum.	jagged leaved.			
9. Bonarienfe.	tree.	Buenos Ayres.	June, Sept.	G. tr.
10. **CAPSICUM**	**CAPSICUM**			
1. annuum.	annual.	Indies.	June, July.	S. an.
2. finenfe.	yellow.	China.		S. pe.
3. groffum.	heart fhaped, or *bell pepper*.	Indies.	June, Aug.	S. pe.
CESTRUM				
1. laurifolium.	laurel-leaved.	America.	Auguft.	S. tr.
2. nocturnum.	night fmelling.	Jamaica, Chili.	November.	S. tr.
3. Parqui.	willow leaved.	Chili.	June, July.	S. tr.
4. diurnum.	day fmelling.	Chili, Havana.	November.	S. tr.
LYCIUM	**BOX-THORN**			
1. afrum.	African.	Africa.	June, July.	G. tr.

Y

PENTANDRIA MONOGYNIA.

SERISSA				
1. foetida.	ftinking.	- Indies, China.	April, July.	G. tr
var. fl. plen.	double flowered.	- Cochinchina, Japan.	April, July.	G. tr.
ARDISIA	ARDISIA, or Aderno.			
1. folanacea.	Conda Mayor.	- Coaft of Coromandel.		S. tr.
2. excelfa.	laurel leaved.	• Madeira.		G. tr.
IACQUINIA	IACQUINIA			
1. armillaris.	obtufe leaved.	• America.		S. tr.
CHIRONIA	CHIRONIA			
1. linoides.	flax-leaved.	- C. G. H.	June, Auguft.	G. tr.
2. baccifera.	berry-bearing.	- Æthiopia.	June, July.	G. tr.
3. frutefcens.	fhrubby.	- ⸺	June, Sept.	G. tr'
CORDIA	CORDIA			
1. Myxa.	fmooth-leaved.	- Egypt, Malabar.		S. tr.
2. Gerafcanthus.	fpear leaved.	- Jamaica.		S. tr.
3. Collococca.	long leaved.	- ⸺		S. tr.
EHRETIA	EHRETIA			
1. tinifolia.	tinus leaved.	- Jamaica.	June, July.	S. tr.
CHRYSOPYLLUM	STAR-APPLE			
1. Cainito.	broad-leaved.	- Martinique.		S. tr.
2. argenteum.	narrow-leaved.	- ⸺		S. tr.

PENTANDRIA MONOGYNIA.

		Habitat	Flowering	
BUMELIA				
1. nigra.		Jamaica.		*S. tr.*
2. tenax.		Carolina.		*G. tr.*
TECTONA	TEAK-WOOD, or *Indian Oak.*			
1. grandis.	great.	Java, Ceylon, Malabar, Coromandel.		*S. tr.*
SIDEROXYLON	IRON-WOOD			
1. inerme.	smooth.	Æthiopia.	July.	*G. tr.*
RHAMNUS	BUCK-THORN			
1. colubrinus.	red wood.	America.	June.	*S. tr.*
2. glandulofus.	Madeira.	Madeira.	June, July.	*G. tr.*
PHYLICA	PHYLICA			
1. ericoides.	heath leaved.	Ethiopia.	Nov., March.	*G. tr.*
2. capitata.	headed.	C. G. H.		*G. tr.*
3. ftipularis.	ftipuled.		March, May.	*G. tr.*
4. buxifolia.	box-leaved.	Ethiopia.	*M. P. S.*	*G. tr.*
5. fpicata.	fpiked.	C. G. H.		*G. tr.*
6. villofa.	villous.			*G. tr.*
7. bicolor.	two coloured.			*G. tr.*
8. eriophoros.	woolly.		August, Nov.	*G. tr.*
9. paniculata.	panicled.			*G. tr.*

Y 2

PENTANDRIA MONOGYNIA.

CEANOTHUS			
1. Afiaticus. —	Ceylon.		*S. tr.*
2. Africanus, —	Æthiopia.	March, April.	*G. tr.*
SCOPOLIA			
1. aculeata. prickly, —	E. Indies.		*S. tr.*
ARDUINA			
1. bifpinofa. two-fpined, —	C. G. H.	March, Auguft,	*G. tr.*
CELASTRUS STAFF-TREE			
1. lucidus. fhining, or fmall Hottentot cherry. —	C. G. H.	*M. P. S.*	*G. tr.*
2. buxifolius, box leaved. —	Ethiopia.	May, June.	*G. tr.*
3. pyracanthus. pyracantha leaved. —	————	*M. P. S.*	*G. tr.*
DIOSMA			
1. oppofitifolia. oppofite-leaved. —	C. G. H.	March, July.	*G. tr.*
2. ericoides. fweet fcented, or heath leav'd. —	————	June, July.	*G. tr.*
3. ciliata. ciliated. —	————	July, Auguft.	*G. tr.*
4. capitata. headed. —	————		*G. tr.*
5. lanceolata. lance leaved. —	————	Auguft.	*G. tr.*
6. latifolia. broad-leaved. —	————	July, Auguft.	*G. tr.*
7. tetragona. four cornered. —	————		*G. tr.*
8. ovata. oval leaved. —	————	May, July.	*G. tr.*
9. uniflora, one flowered, —	————		*G. tr.*

PENTANDRIA MONOGYNIA.

DIOSMA				
10. rubra.	-	C. G. H.	Feb. April.	*G. tr.*
11. marginata.	-	——		*G. tr.*
12. cupressina.	-	——		*G. tr.*
13. ferrulata. fawed leaved.	-	——		
BRUNIA				
1. nudiflora. imbricated.	-	Æthiopia.	*M. P. S.*	*G. tr.*
2. lanuginosa. heath leaved.	-	——		*G. tr.*
3. abrotanoides. thyme leaved.	-	——		*G. tr.*
STAAVIA				
1. glutinosa. clammy.	-	C. G. H.	June, July.	*G. tr.*
PITTOSPORUM				
1. coriaceum. leathery leaved.	-	Canary Iflands.	May, June.	*G. tr.*
2. undulatum. wave leaved.	-	New Holland.	July, Auguft.	*G. tr.*
ITEA				
1. Cyrilla. clufter flowered.	-	Carolina.	June, Auguft.	*G. tr.*
2. fpinofa. fpinous.				
CEDRELA BASTARD CEDAR				
1. odorata. afh leaved.	•	W. India.		*S. tr.*
ELÆODENDRUM				
1. Orientale. willow leaved.	•			*S. tr.*
BILLARDIERA				
1. fcandens. climbing.	•	New Holland.	July, Auguft.	*G. tr.*

PENTANDRIA MONOGYNIA.

LEEA			
1. fambucina.	elder leaved.	E. India.	
HELICONIA	HELICONIA		
1. Bihai.	baftard, or wild plaintain.	-	
2. pfittacorum.	parrot beaked.	S. America.	April, May.	S. tr.
ACHYRANTHES	ACHYRANTHES	Surinam, Jamaica.		S. pe.
				S. pe.
1. lappacea.	burdock feeded.	Indies.	Auguft, Oct.	S. tr.
2. patula.	fpreading.			S. bi.
CELOSIA	CELOSIA			
1. argentea.	filvery.	China.	June, Sept.	S. an.
2. criftata.	coxcomb.	Afia.	July, Sept.	S. an.
3. coccinea.	fcarlet.	Indies.		S. an.
4. caftrenfis.	branched			S. au.
5. trygina.	oval leaved.	Senegal.	Aug. Oct.	S. au.
6. nodiflora.	knotted.	Ceylon.		S. an.
ILLECEBRUM	ILLECEBRUM			
1. feffile.	feffile flowered.	E. India.	July.	S. bi.
RAUWOLFIA	RAUWOLFIA			
1. nitida.	fhining.	S. America.	June, Sept.	S. tr.
CERBERA	CERBERA			
1. Thevetia.	willow leaved.	Cuba, Martinico.		S. tr.

PENTANDRIA MONOGYNIA.

GARDENIA			
1. florida. -	Suratte, Amboina.	June, August.	*G. tr.*
var. fl. plen. double flowered ditto. -	C. G. H.	——	*G. tr.*
2. Thunbergia. starry. -	Manilla, C. G. H.		*S. tr.*
3. spinosa. thorny. -	China.		*S. tr.*
4. dumetorum. spiney. -	E. India.		*S. tr.*
5. Randia. round leaved. -	Jamaica, Martinico.		
ALLAMANDA			
1. cathartica. purging. -	Guiana.	July, August.	*S. tr.*
VINCA PERIWINKLE			
1. rosea. -	Madagascar, Java.	*M. P. S.*	*S. tr.*
var. fl. alb. double flowered ditto. -	——	——	
NERIUM ROSE BAY			
1. Oleander. common Oleander. -	E. India.	June, Oct.	*G. tr.*
var. fl. plen. double flowered ditto. -	——	——	
2. coronarium. broad leaved. -	——		*S. tr.*
TABERNAEMONTANA. TABERNAEMONTANA.			
1. citrifolia. citron leaved American. -	America.		*S. tr.*

PENTANDRIA DIGYNIA.

DIGYNIA.

PERGULARIA sweet smelling.			
1. odoratissima.	E. India.	May, August.	*S. tr.*
PERIPLOCA			
1. laevigata. smooth.	Canary Islands.	July.	*G. tr.*
2. Africana. African.	C. G. H.	June, Sept.	*G. tr.*
CYNANCHUM			
1. erectum. upright.	Syria.		*G. tr.*
ASCLEPIAS SWALLOW-WORT			
1. Curassavia. Curassavian.	Curasa.	June, Sept.	*S. tr*
2. parviflora. small flowered.	Carolina, E. Florida.	July, Sept.	*G. pe.*
3. arborescens. tree.	C. G. H.	Oct. Dec.	*G. tr.*
4. fruticosa. willow leaved, or shrubby.		June, Sept.	*G. tr.*
5. gigantea. curled flowered gigantic.	Indies.	July, Sept.	*S. tr.*
MELODINUS			
1. scandens. climbing.	New Holland.		*G. tr.*
STAPELIA			
1. revoluta. revolute.	C. G. H.	May, Nov.	*S. tr.*
2. hirsuta. hairy.	——	——	*S. sr.*
3. grandiflora. great flowered.	——	——	*S. tr.*
4. acuminata. acuminated.	——	——	*S. tr.*

PENTANDRIA DIGYNIA.

STAPELIA	STAPELIA			
6. caefpitofa.	turfy.	C. G. H.	May, Nov.	S. tr.
7. pedunculata.	long ftalked.	———		S. tr.
8. variegata.	variegated.	———		S. tr.
CHENOPODIUM	GOOSE-FOOT			
1. Guineenfe.	Guinea.	Guinea.	July, September.	G. an.
SALSOLA	SALT-WORT			
1. proftrata.	trailing.	Afia, Spain, Auftria, Siberia.		G. tr.
BOSEA	GOLDEN ROD TREE			
1. Yerramora.	Canary.	Canary Iflands.		G. tr.
GENTIANA	GENTIAN			
1. Catefbaei.	foap-wort.	Virginia, Penfylvania.	Auguft, Sept.	G. tr.
PHYLLIS	BASTARD HARE'S EAR			
1. nobla.	Canary.	Canary Iflands.	June, July.	G. tr.
FALKIA	FALEIA			
1. repens.	creeping.	C. G. H.	May, Auguft.	G. pe.
BUBON	BUBON			
1. Macedonicum.	Macedonian Parfly.	Macedonia, Mauritiana.	June Auguft.	G. pe.
2. gummiferum.	gum bearing.	Æthiopia.	July.	G. tr.
var.fl.rub.	red flowered ditto.	———		

Z

PETANDRIA TRIGYNIA.

RHUS	SUMACK			
1. fuccedaneum.	red lac.	Japan, China.		G. tr.
2. incifum.	cut leaved.	C. G. H.		G. tr.
3. tomentofum.	woolly leaved.			G. tr.
4. viminale.	willow leaved.			G. tr.
5. anguftifolium.	narrow leaved.	Æthiopia.		G. tr.
6. laevigatum.	fmooth leaved.	C. G. H.		G. tr.
7. lucidum.	great fhining leaved			G. tr.
var. minor,	fmall ditto.			G. tr.
8. cuneifolium.	wedge leaved.			G. tr.
CASSINE	CASSINE			
1. Capenfis.	Cape.	C. G. H.	July, Auguft.	G. tr.
2. Maurocenia.	Hottentot cherry.	Æthiopia.		G. tr.
3. media.				
SPATHELIA	SPATHELIA			
1. fimplex.	fumack leaved.	Jamaica.		S. tr.
XYLOPHYLLA	XYLOPHYLLA			
1. latifolia.	broad leaved.	Surinam, Jamaica.	Auguft, Oct.	S. tr.
2. anguftifolia.	narrow leaved.	Bahama Iflands.	July, Aug.	S. pe.
TURNERA	TURNERA			
1. ulmifolia.	elm leaved.	Jamaica.		S. tr.

PENTANDRIA TRYGYNIA

PHARNACEUM				
1. incanum.	linear leaved.	Africa.	*M. P. S.*	*G. pe.*
BASELLA	BASELLA, or Malabar Nightshade.			
1. rubra.	red.	Lima.	July, Nov.	*S. bi.*
2. alba.	white.	China, Amboina.		*S. bi.*
PORTULACARIA	PUUSLANE TREE			
1. Afra.	African.	Africa.	——	*G. tr.*

PENTAGYNIA.

STATICE	THRIFT			
1. suffruticosa.	narrow leaved shrubby.	Siberia.		*G. tr.*
2. monopetala.	broad leaved.	Sicily.		*G. tr.*
3. ferulacea.		Barbary, Portugal, Spain.	July, Aug.	*G. pe.*
4. sinuata.	scollop leaved.	Sicily, Palestine, Africa.		*G. pe.*
5. mucronata.	curled-stalked.	Barbary.	*M. P. Y.*	*G pe.*
LINUM	FLAX			
1. Africanum.	shrubby	C. G. H.	July, Aug.	*G. tr.*
CRASSULA				
1. coccinea.	scarlet flowered.	Æthiopia.	June, Aug.	*G. tr.*
2. scabra.	rough leaved.	C. G. H.	July, Aug.	*G. tr.*
3. perfoliata.	perfoliate.	Æthiopia.	——	*G. tr.*

Z 2

PENTANDRIA PENTAGYNIA.

CRASSULA	CRASSULA			
4. mollis.	soft leaved.	Æthiopia.	August.	G. tr.
5. tetragona.	four sided.		July, August.	G. tr.
6. cultrata.	sharp leaved.		June.	G. tr.
7. imbricata.	imbricated.	C. G. H.	April, May.	G. tr.
8. obliqua.	oblique leaved.		May, August.	G. tr.
9. cordata.	heart leaved.		Sept. Oct.	G. tr.
10. lactea.	white.			G. tr.
11. arborescens.	tree.	Æthiopia.	July, August.	G. tr.
12. ciliata.	ciliated.	C. G. H.	May, August.	G. tr.
13. marginalis.	marginated.		March, June.	G. pt.
14. odoratissima.	sweet smelling.			G. tr.
MAHERNIA	MAHERNIA			
1. pinnata.	winged.	C. G. H	June, August.	G. tr.
2. incisa.	cut leaved.		June, July.	G. tr.

CLASSIS VI.
HEXANDRIA MONOGYNIA.

PITCAIRNIA	PITCAIRNIA			
1. bromeliaefolia.	pine apple leaved.	Jamaica.	June.	S. tr.

HEXANDRIA MONOGYNIA.

PITCAIRNIA			
2. angustifolia. narrow leaved.	Santa Cruz.	Dec. Jan.	S. tr.
TRADESCANTIA SPIDER WORT			
1. erecta. erect.	Mexico.		G. an.
2. Zanonia. gentian leaved.	Jamaica, Guinea.	July, Sept.	S. pe.
3. discolor. two coloured.	S. America.	M. P. S.	S. pe.
HAEMANTHUS BLOOD FLOWER			
1. coccineus. scarlet.	C. G. H.	August, Oct.	G. pe.
2. puniceus. colchicum leaved.		May, July.	G. pe.
3. multiflorus. many flowered.	Sierra Leone.	May, Sept.	G. pe.
4. ciliaris. ciliated.	C. G. H.		G. pe.
5. coarctatus. compressed.		Sept.	G. pe.
TULBAGIA			
1. alliacea. narcissus leaved.	C. G. H.	June, August.	G. pe.
PANCRATIUM			
1. Caribaeum. Caribean.	Jamaica, W. India.	M. P. S.	S. pe.
2. Carolinianum. Carolina.	Carolina.		S. pe.
3. littorale. tall.	W. India.	M. P. S.	S. pe.
4. speciosum. shewy.	E. India.		S. pe.
5. Amboinense. broad leaved.	Amboina.		S. pe.
CRINUM			
1. Asiaticum. Asiatic.	Malabar, Ceylon, S. America.		S. pe.
2. Americanum. great American.	S. America.	Var. Seasons.	S. pe.

HEXANDRIA MONOGYNIA.

			Var. Seasons.	
CRINUM				
3. erubescens.	small American.	W. India.		S. pe.
AGAPANTHUS	AGAPANTHUS, or *blue African lily*			G. pe.
1. umbellatus.	umbelled.	C. G. H.	July, August.	G. pe.
CYRTANTHUS	CYRTANTHUS		M. P. S.	G. pe.
1. angustifolius.	narrow leaved.	C. G. H.		
AMARYLLIS	AMARYLLIS			
1. formosissima.	Jacobean lily.	S. America.	May, June.	S. pe.
2. reginæ.	Mexican lily.			S. pe.
3. linearis.	linear leaved.	C. G. H.	April, March.	G. pe.
4. equestris.	Barbadoes lily.	W. India.		S. pe.
5. vittata.	ribbon.	C. G. H.	July.	G. pe.
6. falcata.	fickle leaved.		June.	G. pe.
7. ornata.	Cape coast lily.	Guinea.	June, July.	S. pe.
8. longifolia.	long leaved.	C. G. H.	July.	G. pe.
9. aurea.	golden.	China.	August, Jan.	S. pe.
10. undulata.	waved leaved.	C. G. H.	March, April.	G. pe.
11. crispa.	curled flowered.			G. pe.
12. Jagus.	large.	Guinea.	June.	G. pe.
ALLIUM	GARLICK			S. pe.
1. gracile.	Jamaica.	Jamaica.		
LILIUM	LILY			S. pe.
1. cordifolium.	heart leaved.	Japan.	May, June.	G. pe.

HEXANDRIA MONOGYNIA.

EUCOMIS.				
1. regia.	tongue leaved.	C. G. H.	March, May.	*G. pe.*
2. undulata.	waved leaved.	——	——	*G. pe.*
3. punctata.	spotted.	——	July.	*G. pe.*
GLORIOSA				
1. superba.	superb lily.	East India, Malabar.	July, August.	*S. pe.*
ALBUCA				
1. major.	great.	C. G. H.	May	*G. pe.*
2. minor.	small.	——	May, June.	*G. pe.*
HYPOXIS				
1. villosa.	villose.	C. G. H.	May.	*G. po.*
2. plicata.	plaited leaved.	——	——	*G. pe.*
ORNITHOGALUM STAR OF BETHLEHEM				
1. Arabicum.	Arabian.	Alexandria, Madeira.	March, April.	*G. pe.*
2. aureum.	golden.	C. G. H.	——	*G. pe.*
3. caudatum.	long spiked.	——	March, August.	*G. pe.*
DIANELLA				
1. caerulea.	blue.	C. G. H.	May, June.	*G. tr.*
ANTHERICUM				
1. revolutum.	curled flowered.	C. G. H.	Sept.	*G. rt.*
2. frutescens.	shrubby.	——	*M. P. S.*	*G. tr.*
3. aloides.	aloe leaved.	——	——	*G. pe.*

HEXANDRIA MONOGYNIA.

ASPARAGUS	ASPARAGUS			
1. retrofractus.	larch leaved.	Africa.	Aug. Sept.	*G. tr.*
2. albus.	white.	Spain, Portugal.		*G. tr.*
3. aphyllus.	prickly.	Sicily, Spain, Portugal.		*G. tr.*
4. farmentofus.	creeping rooted.	Ceylon.	Auguft.	*G. tr.*
DRACÆNA	DRACÆNA			
1. Draco.	dragon.	E. India.		*S. tr.*
2. cernua.	drooping leaved.	Mauritius.		*S. tr.*
3. ferrea.	purple leaved.	China.	March, April.	*S. tr.*
4. marginata.	aloe leaved	Bourbon.	April.	*S. tr.*
5. enfifolia.	fword leaved.	E. India.	Auguft.	*S. pe.*
POLYANTHES	TUBEROSE			
1. tuberofa	common	Java, Ceylon.	Aug. Sept.	*G. pe.*
PHORMIUM	FLAX LILY			
1. tenax.	New Holland	New Holland.		*G. pe.*
LACHENALIA	LACHENALIA			
1. viridis.	green flowered.	C. G. H.		*G. pe.*
2. tricolor.	three coloured.	———	April,	*G. pe.*
3. pendula.	pendulous.	———	March, April.	*G. pe.*
4. orchioides.	fpotted leaved	———	———	*G. pe.*
5. luteola.				*G. pe.*
VELTHEIMIA	VELTHEIMIA			
1. viridifolia.	green leaved,	C. G. H.		*G. pe.*

HEXANDRIA MONOGYNIA.

VELTHEIMIA				
2. glauca.	glaucous.	C G H.	January.	G. p.
3. uvaria.	orange flowered	———	July, August.	G. p.
ALETRIS				
1. fragrans.	sweet scented.			
2. farmentofa.	creeping rooted.	Africa.	Feb. March.	S. tr.
YUCCA	DEVIL'S NEEDLE	C. G. H.	Oct. Jan.	G. p.
1. aloifolia.	aloe leaved.	Jamaica, Vera Cruce.	August, Sept.	G. tr.
ALOE				
1. arborefcens.	tree.	Africa.	V. feafons.	S. tr.
2. fuccotrina.	fuccotrine.	———		S. tr.
3. faponaria.	great foap.	———		S. tr.
4. obfcura.	common foap.	———		S. tr.
5. mitriformis	great mitre.	———		S. tr.
6. brevifolia.	fmall mitre.	———		S tr.
7. arachnoides.	common cobweb.	C. G. H.		S. tr.
var. pumil.	fmall ditto.	———		S. tr.
8. margaritifera.	pearl.	———		S. tr.
var. major.	great ditto.	———		S. tr.
var. minor.	fmall ditto.	———		S. tr.
9. verrucofa.	warted.	———		S. tr.
10. carinata.	keel leaved.	Africa.		S. tr.

A 2

HEXANDRIA MONOGYNIA.

ALOE				
11. maculata.	narrow leaved spotted.	C. G. H.	V. seasons.	S. tr.
var. *obliqua.*	*broad ditto.*	⎸	⎸	S. tr.
12. lingua.	tongue.	⎸	⎸	S. tr.
13. crassifolia.	thick tongue.	⎸	⎸	S. tr.
14. angustifolia.	narrow tongue.	⎸	⎸	S. tr.
15. sinuata.	sinuated.	Barbadoes, C. G. H.	⎸	S. tr.
16. plicatilis.	fan.	Africa.	⎸	S. tr.
17. variegata.	partridge breast.	Æthiopia.	⎸	S. tr.
18. viscosa.	upright triangular.	C. G. H.	⎸	S. tr.
19. spiralis.	spiral.	Africa.	⎸	S. tr.
20. retusa.	cushion.	⎸	⎸	S. tr.
AGAVE	AGAVE			
1. Americana.	American.	S. America.	V. seasons.	S. tr.
var. fol. *varieg.*	*variegated ditto.*	⎸	⎸	
ALSTRŒMERIA	ALSTRŒMERIA			
1. Pelegrina.	spotted flowered.	Peru, Lima.	June, Sept.	G. pr.
2. Ligtu.	striped flowered.	Lima.	Feb. March.	S. pr.
HEMEROCALLIS	DAY-LILY			
1. Japonica.	blue flowered	Japan.	May, June.	G. pr.
CALAMUS	CALAMUS			
1. Rotang.	Rotang reed.	E. India.		S. tr.

HEXANDRIA MONOGYNIA.

RICHARDIA			
1. scabra,	rough.	Vera Cruce.	S. tr.
ACHRAS			
1. mammosa,	Mammee sapota.	Cuba, Jamaica, Carthagena.	S. tr.
2. Sapota.	common Sapota.	S. America.	S. tr.
PRINOS	WINTER BERRY		
1. lucidus.	shining.		June, July.	G. tr.
CANARINA			
1. campanula.	Canary bell flower.	Canary Islands.	Feb. April.	G. pe.
BAMBUSA	BAMBOO CANE		
1. arundinacea.	reed.	Both Indies.	S. tr.

DIGYNIA.

ATRAPHAXIS			
1. undulata.	waved leaved.	C. G. H.	June, July.	G. tr.

TRIGYNIA.

RUMEX	DOCK		
1. Lunaria.	tree sorrel.	Canary Islands.	July, August.	G. tr.

A 2 2

HEXANDRIA TRIGYNIA.

FLAGELLARIA
1. Indica.

FLAGELLARIA
Indian. - Java, Malabar, Ceylon, Guinea.   *S. tr,*

MEDEOLA
1. asparagoides.
2. angustifolia.

MEDEOLA
broad leaved shrubby. Æthiopia. Oct. March. *G. tr.*
narrow leaved. - C. G. H. *G. tr.*

TETRAGYNIA.

PETIVERIA
1. alliacea.

GUINEA-HEN-WEED
common. - Jamaica. *M. P. S.* *S. tr.*

CLASSIS. VII.

HEPTANDRIA MONOGYNIA.

DISANDRA
1. prostrata.

DISANDRA.
trailing Madeira. *M. P. S.* *G. p.*

HEPTAGYNIA.

	SEPTAS Cape.	C. G. H.	August, Sept.	G. p.
SEPTAS 1. Capensis.				

CLASSIS VIII.
OCTANDRIA MONOGYNIA.

	OENOTHERA	C. G. H.	June, August.	G. tr.
OENOTHERA 1. nocturna.	Cape.			
2. odorata.	sweet scented.	Patagonia.	M. P. S.	G. p.
3. rosea.	rose flowered.	Peru.		G. p.
GAURA 1. mutabilis.	GAURA changeable.	S. America.	July.	G. bi.
CORREA 1. alba.	CORREA white.	New Holland.	April, June.	G. tr.
XIMENIA 1. Americana.	XIMENIA American.	America.		S. tr.
FUCHSIA 1. coccinea.	FUCHSIA scarlet.	Chili.	May, August.	G. tr.
DODONAEA 1. viscosa.	DODONAEA viscous.	Indies.	July.	S. tr.
2. triquetra.	three sided.	New Holland.	July.	G. tr.

OCTANDRIA MONOGYNIA.

VACCINIUM				
1. Arctostaphyllos.	WHORTLE-BERRY. Madeira.	Cappadocia, Madeira.	June, July.	G. tr.
ERICA	HEATH			
1. Halicacaba.	Alkakengi flowered.	C G. H.	May, July.	G. tr.
2. discolor.	two coloured.	—	May, June.	G. tr.
var. *concolor.*			*M. P. S.*	
3. cruenta.	bloody.	—	May, June.	G. tr.
4. nigrita.	larch.	—	May, June.	G. tr.
5. regerminans.	sprouting.	—	May, June.	G. tr.
6. urceolaris.	pitcher flowered.	—		G. tr.
7. marifolia.	marum leaved.	—		G. tr.
8. pubescens.	downy.	—	Feb. Jane.	G. tr.
9. perfoluta.	garland.	—	Mar. May.	G. tr.
10. caffra.	Caffre.	—	Mar. June.	G. er.
11. mucosa.	mucous.	—	*M. P. S.*	G. tr.
12. abietina.	fir.	—		
13. verticillata.	whorl flowered.	—	*M. P. S.*	G. tr.
14. mammosa.	nipple flowered.	—	Feb. July.	G. tr.
15. empetrifolia.	crow berry leaved.	—	April, May.	G. tr.
16. spicata.	spiked.	—	May, June.	G. tr.
17. sessiliflora.	sessile flowered.	—	June, August.	G. tr.
18. fascicularis.	crown flowered.	—		G. tr.

OCTANDRIA MONOGYNIA.

ERICA				
	HEATH			
19. corifolia.	coris-leaved.	C. G. H.	May, Sept.	G. tr.
20. triflora.	three flowered.	S. of Europe.	April, May.	G. tr.
21. fcoparia.	fmall green flowered	———		G. tr.
22. Auftralis.	Spanifh.			G. tr.
23. ramentacea.	long flender branched.	C. G. H.	June, July.	G. tr.
24. imbecilla.	feeble.	———		G. tr.
25. margaritacea.	pearl flowered.	———	May, July.	G. tr.
26. baccans.	arbutus flowered.	———	April, May.	G. tr.
27. pendula.	pendulous.	———	July, Auguft.	G. tr.
28. phyfodes.	bladder.	———	March, June.	G. tr.
29. cernua.	drooping flowered.	Portugal.	June, Sept.	G. tr.
30. umbellata.	umbelled.	C. G. H.	May, July.	G. tr.
31. nudiflora.	naked flowered.	———	June, Auguft.	G. tr.
32. fucata.	painted.	C. G. H.	May, July.	G. tr.
33. taxifolia.	yew leaved.	———	July, Sept.	G. tr.
34. capitata.	woolly-calyxed.	———	May, Auguft.	G. tr.
35. Petiverii.	Petiver's.	———	March, July.	G. tr.
36. Petiveriana.	Petiverian.	———		G. tr.
37. hifpidula.	roughifh.	———	July, Auguft.	G. tr.
38. Banksii.	white Bank's.	———	May, June.	G. tr.
var. purp.	purple ditto.	———	———	

OCTANDRIA MONOGYNIA.

	HEATH	C. G. H.		G. tr.
ERICA				
39. Sebana.	Seba's yellow.	•	Mar. June.	G. tr.
var. *fusca.*	*brown ditto.*	|		G. tr.
— *virid.*	*green ditto.*	|		G. tr.
40. flexuosa.	zigzag.	|		G. tr.
41. monadelphia.	monadelphous.	|	June, July.	G. tr.
42. Plukenetii.	Plukenets.	|	April, May.	G. tr.
43. versicolor.	various coloured.	|	Mar. May,	G. tr.
44. costata.	rib-flowered.	|	|	G. tr.
45. pulchella.	neat.	|		G. tr.
46. longifolia.	long leaved.	|	May, July.	G. tr.
47. coccinea.	scarlet.	|	*M. P. S.*	G. tr.
48. vestita.	close leaved purple.	|	V. seasons.	G. tr.
49. purpurea.	purple.	|	April, June.	G. tr.
50. concinna.	blush flowered.	|	Sept. Oct.	G. tr.
51. grandiflora.	large flowered.	|	May, July.	G. tr.
52. curviflora.	curve flowered.	|	August, Dec.	G. tr.
53. simpliciflora.	single flowered.	|	March, June.	S. tr.
54. ignescens.	fiery-flowered.	|	June, July.	G. tr.
55. tubiflora.	tube-flowered.	•	April, June.	G. tr.
56. procera.	tall.	•	April, July.	G. tr.
57. conspicua.	long-tubed yellow.	•	May, August.	G. tr.

OCTANDRIA MONOGYNIA.

ERICA	HEATH	C.G.H.	M.P.S.	
58. cerinthoides.	honey wort.	—		G. tr.
var. elatior.	tall ditto.	—		G. tr.
59. Massoni.	Masson's.	—	May, August.	G. tr.
60. ventricosa.	bellied.	—		G. tr.
61. Walkeri.	Walker's.	—		G. tr.
62. comosa.	tufted.	—	April, May.	G. tr.
63. denticulata.	white toothed.	—	June, July.	G. tr.
64. muscosa.	musk.	—	June, July.	G. tr.
65. viscaria.	clammy.	—	June, Sept.	G. tr.
66. mollissima.	very soft.	—	April, May.	G. tr.
67. declinata.	bending stalked.	—	April, May.	G. tr.
68. Sparrmanni.	Sparrmans.	—	August, Sept.	G. tr.
69. divaricata.	spreading.	—	May, July.	G. tr.
70. hirtiflora.	hairy flowered.	—		G. tr.
71. Hibertia.	Hibert's.	—		G. tr.
72. tenella.	delicate.	—		G. tr.
73. filamentosa.	thready.	—		G. tr.
74. nivea.	snowy.	—		G. tr.
75. Blearia.	Blearias.	—		G. tr.
76. Linnæa.	Linnæus's.	—		G. tr.
77. gelida.	icy.	—		G. tr.

B b

OCTANDRIA MONOGYNIA.

ERICA	HEATH			
78. lævis.	smooth.	C. G. H.		G. tr.
79. foliosa.	leafy.	——		G. tr.
80. virgata.	twiggy.			G. tr.
DAPHNE	DAPHNE			
1. odora.	sweet scented.	China, Japan.	Jan. March.	G. tr.
2. collina.	hairy.	Italy.	March, April.	G. tr.
GNIDIA	GNIDIA			
1. pinifolia.	pine leaved.	C. G. H.	June, August.	G. tr.
2. simplex.	flax leaved.	C. G. H.	M. P. S.	G. tr.
GRISLEA	GRISLEA			
1. tomentosa.	downy.	S. America.		S. tr.
PASSERINA	AFRICAN SPARROW WORT			
1. filiformis.	heath leaved.	Æthiopia.	June, July.	G. tr.
2. grandiflora.	large flowered.	Africa.	May, June.	G. tr.
3. laxa.	loose.	C. G. H.		G. tr.
4. spicata.	spiked.	——		G. tr.
LACHNÆA	LACHNÆA			
1. conglomerata.	cluster headed.	C. G. H.	June, July.	G. tr.

DIGYNIA.

GALENIA	GALENIA African.	C. G. H.	June, August.	G. tr.
1. Africana.				

TRIGYNIA.

COCCOLOBA	SEA-SIDE-GRAPE.			
1. uvifera.	round leaved.	America, W. India.		S. tr.
2. pubescens.	downy.	—		S. tr.
3. excoriata.	oval leaved.	—		S. tr.
4. emarginata.	emarginated.	S. America.		S. tr.
5. flavescens.	yellowish.	Domingo.		S. tr.
CARDIOSPERMUM.	HEART-SEED.			
1. Halicacabum.	smooth leaved.	Indies.	July, Aug.	S. an.
SAPINDUS	SOAP BERRY			
1. saponaria.	common.	S. America.		S. tr.

TETRAGYNIA.

HALORAGIS	HALORAGIS			
1. Cercodia.	whorled flowered.	New Holland.	M. P. S.	G. tr.

B b 2

C L A S S I S   IX.

ENNEANDRIA MONOGYNIA.

LAURUS	LAUREL			
1. Cinnamomum.	cinnamon.	Ceylon, America, Martinique, Jamaica.		*S. tr.*
2. Cassia.	bastard cinnamon.	Malabar, Java, Sumatra.	July.	*S. tr.*
3. Camphora.	camphire.	Japan.		*G. tr.*
4. Indica.	royal Bay.	Madeira.	October.	*G. tr.*
5. foetens.	Madeira, or Til.	Canary Islands, Madeira.		*G. tr.*
6. Borbonia.	broad leaved bay.	Carolina, Virginia.	April, May.	*G. tr.*
7. Martinicensis.	Martinicoe.	Martinique.		*S. tr.*
ANACARDIUM	CASHEW NUT.			
1. Occidentale.	common.	Indies.		*S. tr.*

C L A S S I S   X.

DECANDRIA MONOGYNIA.

SOPHORA	SOPHORA		
Occidentalis.	Occidenal.	America.	*S. tr.*

DECANDRIA MONOGYNIA.

PODALYRIA				
1. Capensis.	Cape.	C G. H.	July.	G. tr.
2. biflora	two flowered.	———	Jan. May.	G. tr.
3. myrtillifolia.	myrtle leaved.	———		G. tr.
4. hirsuta.	hairy.	———	May, June.	G. tr.
5. aurea.	golden.	Abyssinia.		S. tr.
PULTENAEA				
1. linophylla.	flax leaved.	New Holland.	May, July.	G. tr.
2. daphnoides.	daphne.	———	June, July.	G. tr.
3. stipularis.	heath leaved.	———	May, July.	G. tr.
4. villosa.	villous.	———	April, May.	G. tr.
5. paleacea.	chaffy.	———		G. tr.
6. juncea.	rush leaved.	———		G. tr.
BAUHINIA, or *Mountain Ebony*				
1. porrecta.	smooth leaved.	Jamaica, Hispaniola.	July.	S. tr.
2. candida.	downy leaved.	E. India.		S. tr.
HYMENAEA				
1. Courbaril	locust tree.	W. India.		S. tr.
PARKINSONIA				
1. aculeata.	prickly.	W. India.		S. tr.
CASSIA				
1. fistula	purging.	India, Egypt.		S. tr.
2. biflora.	two-flowered.	America.	July, August.	S. tr.

DECANDRIA MONOGYNIA.

CASSIA				
3. multiglandulosa.	glandulous.	Teneriffe.	July, Aug.	S. tr.
4. hirsuta.	hairy.	America.		S. tr.
5. ligustrina.	privet leaved.	Bahama Islands.	July.	S. tr.
6. arborescens.	arborescent.	Egypt.		S. tr.
7. alata.	winged.	S. America.		S. tr.
8. mollis.	soft.			S. tr.
CAESALPINIA	CÆSALPINIA, or *Brasileto*.			
1. bijuga.	broad leaved.	Jamaica.		S. tr.
2. pulcherrima.	splendid.	Indies.		S. tr.
3. elata.	lofty.	E. India.		S. tr.
4. Brasilensis.	smooth.	Carolina, Jamaica, Brazil.	July, Oct.	S. tr.
5. Sappan.	prickly leaved.	Indies.		S. tr.
6. mimosoides.	mimosa leaved.	E. India.		S. tr.
GUILANDINA	BONDUC, or *Nicker tree*.			
1. Bonduc.	yellow seeded.	Indies.		S. tr.
2. Bonducella.	grey seeded.			S. tr.
HYPERNANTHERA	HYPERNANTHERA			
1. Moringa.	smooth.	Ceylon, America, Egypt.		S. tr.
SCHOTIA	SCHOTIA			
1. speciosa.	lentiscus leaved.	Senegal, C. G. H.		G. tr.
GUAIACUM	GUAIACUM, or *Lignum vitæ*			
1. officinale.	officinal.	Hispaniola, Jamaica.		S. tr.

DECANDRIA MONOGYNIA.

RUTA	RUE			
1. Chalpenfis.	Aleppo.	- Arabia.	Aug. Sept.	G. tr.
HAEMATOXYLON	LOGWOOD			
1. Campechianum.	common.	- S. America.		S. tr.
MURRAYA	MURRAYA			
1. exotica.	afh leaved.	- E. India.	Auguft, Sept.	S. tr.
TRICHILIA	TRICHILIA			
1. glabra.	fmooth.	- Havanna.		S. tr.
SWIETENIA	MAHOGANY TREE			
1. Mahagoni.	common.	- W. India.		S. tr.
MELIA	BEAD-TREE			
1. Azederach.	common.	- Syria, Ceylon.	June, Auguft.	G. tr.
2. Azedirachta.	Indian.	- E. India.		S. tr.
ZYGOPHYLLUM	BEAN-CAPER			
1. Morgfana.	four leaved.	- C. G. H.	Auguft, Sept.	G. tr.
2. arboreum.	tree.	- S. America.		S. tr.
3. fpinofum.	fpinous.	- C. G. H.		G. tr.
FAGONIA	FAGONIA			
1. Cretica.	Cretan.	- Crete.	July, Auguft.	G. bi.
LIMONIA	LIMONIA			
1. pentaphylla.	five leaved.	- E. India.		S. tr.
CLETHRA	CLETHRA			
1. arborea.	tree.	- Madeira.	July, Auguft.	G. tr.

DECANDRIA MONOGYNIA.

CROWEA					
1. saligna.	willow leaved.	-	New Holland.	July, Nov.	*G. tr.*

DIGYNIA.

ROYENA, or *African bladder nut.*					
1. lucida.	shining.	-	C. G. H.	May, June.	*G. tr.*
2. villosa.	villous.	-	———		*G. tr.*
3. polyandra.	broad leaved.	-	———		*G. sr.*
4. hirsuta.	hairy.	-	———		*G. sr.*

TRIGYNIA.

SILENE	CATCHFLY				
1. gigantea.	gigantic.	-	Africa.	June, July.	*G. bi.*
2. ornata.	dark flowered.	-	C. G. H.	May, August.	*G. bi.*
3. fruticosa.	shrubby.	-	Sicily.	June, July.	*G. tr.*
BRUNNICHIA	BRUNNICHIA				
1. cirrhosa.	tendriled.	-	Bahama Islands.		*S. tr.*
MALPIGHIA	MALPIGHIA, or *Barbadoes-cherry.*				
1. glabra.	smooth,	-	Jamaica, Brazil, Surinam, Curassoa.	Feb. Mar.	*S. tr.*

DECANDRIA TRIGYNIA.

MALPIGHIA, or *Barbadoes-cherry.*				
2. angustifolia.	narrow leaved.	-	W. India.	S. tr.
3. coccifera.	holly leaved.	-	S. America.	S. tr.
BANISTERIA				
1. purpurea.	purple.	-	W. India.	S. tr.
2. ovata.	oval leaved.	-	St. Domingo.	S. tr.

PENTAGYNIA.

AVERRHOA					
1. Bilimbi.	ash leaved.	-	E. India.	May, June.	S. tr.
SPONDIAS	HOG-PLUMB				
1. Myrobalanus.	yellow.	-	W. India.		S. tr.
COTYLEDON	NAVEL-WORT				
1. orbiculata.	round leaved.	-	C. G. H.	July, August.	G. tr.
2. faſcicularis,	cluſter leaved.	-	————	July, Auguſt.	G. tr.
3 ſpuria.	narrow leaved.	-	————		G. tr.
SEDUM	STONE-CROP				
1. nudum.	naked branched.	-	Madeira.		G. pe.
OXALIS	WOOD SORREL				
1. monophylla.	ſingle leaved:	-	C. G. H.	Mar. June.	G. pe.
2. purpurea.	purple.	-	————	Jan. Mar.	G. pe.

C c

DECANDRIA PENTAGYNIA.

OXALIS	WOOD SORREL			
3. tetrapylla.	four leaved.	- Mexico.	May, June.	G. pe.
4. violacea.	violet.	- Virginià, Canada.		G. pe.
5. cernua.	drooping.	- C. G. H.	March, June.	G. pe.
6. verficolor.	various coloured.	- ———	———	G. pe.
7. hirta.	hairy.	- ———	———	G. pe.
var. *fl. plen.*	*double flowered ditto.*	- ———	———	|
LYCHNIS				
1. coronata.	Chinefe.	- China, Japan.	June, July.	G. pe.

DECAGYNIA.

PHYTOLACCA				
1. icofandra.	red.	- E. India.	July, Sept.	S. pe.

CLASSIS XI.

DODECANDRIA MONOGYNIA.

BOCCONIA, or *Tree Celandine.*				
1. frutefcens.	fhrubby.	{ Mexico, Jamaica, Cuba, Domingo. }	Feb. Apr.	S. tr

DODECANDRIA MONOGYNIA.

BOCCONIA	BOCCONIA, or *Tree Celandine.*			
2. cordata.	heart leaved.	- China.	May, July.	*G. p.*
DECUMARIA	DECUMARIA			
1. farmentofa.	farmentofe.	- Carolina.		*G. tr.*
CRATÆVA	CRATÆVA			
1. Tapia.	Tapia.	- Indies.		*G. tr.*
PORTULACA	PURSLANE			
1. pilofa.	hairy.	- W. India.		*S. an.*
TALINUM	TALINUM			
1. triangulare.	triangular.	- America.		*S. tr.*
LYTHRUM	WILLOW-HERB			
1. racemofum.	clustered.	- S. America.		*S. tr.*
STERCULIA	STERCULIA			
1. Balanghas.	laurel leaved.	- Indies.	June, July.	*S. tr.*

TRIGYNIA.

EUPHORBIA	SPURGE			
1. officinarum.	officinal.	- Æthiopia, Africa.	June, July.	*S. tr.*
2. meloformis.	melon.	- C. G. H.	May, Sept.	*S. tr.*
3. Caput-Medufæ.	Medufa's head.	- Æthiopia.	May, Sept.	*S. tr.*
4. Tithymaloides.	laurel leaved.	- S. America.	June, July.	*S. tr.*
5. mellifera.	honey-bearing.	- Madeira.	April, May.	*G. tr.*

TRIGYNIA.

EUPHORBIA	SPURGE			
6. punicea.	scarlet flowered.	Jamaica.	-	Jan. May.	S. tr.
7. sylvatica.	wood.	S. Europe.	-		G. pe.
8. hystrix.	porcupine.	C. G. H.	-	June, July.	S. tr.
9. nereifolia.	oleander leaved.	India.	-	June, July.	S. tr.
SEMPERVIVUM	HOUSE-LEEK				
1. arboreum.	tree.	Portugal, Crete.	-	March.	G. tr.
var. fol. varieg.	variegated do.				
2. Canariense.	Canary.	Canary Islands.	-	June, July.	G. tr.
3. monanthos.	clustered.			July.	G. pe.

CLASSIS XII.

ICOSANDRIA MONOGYNIA.

CACTUS	CACTUS			
1. mammillaris.	small mellon thistle.	S. America.	July, August.	S. tr.
2. nobilis.	15 angled torch thistle.	Mexico.		S. tr.
3. heptagona.	seven angled torch thistle.	America.		S. tr.
4. tetragona.	four angled torch thistle.	S. America.	July.	S. tr.
5. pentagona.	five angled torch thistle.	America.		S. tr.

ICOSANDRIA MONOGYNIA.

CACTUS				
6. repanda.	waved torch thistle.	S. America.	August.	S. tr.
7. lanuginosa.	woolly.	Curassao.	—	S. tr.
8. grandiflora.	night flow. cereus.	Jamaica, Vera Cruce.	July, August.	S. tr.
9. flagelliformis.	small creeping cereus.	S. America.	March, June.	S. tr.
10. Opuntia.	Indian fig.	America, Peru, Virginia, Spain, Italy, Portugal, Minorca.	July, August.	S. tr.
11. coccinellifer.	cochineal Ind fig.	Jamaica, S. America.	—	S. tr.
LEPTOSPERMUM				
1. scoparium.	myrtle leaved.	New Zealand.	June, July.	G. tr.
2. Thea.	Botany bay tea.	New Holland.	—	G. tr.
3. flavescens.	yellowish.	—	—	G. tr.
4. pubescens.	downy.	—	—	G. tr.
5. parvifolium.	small leaved.	—	June, July.	G. tr.
6. baccatum.	berry bearing.	—	—	G. tr.
7. lanigerum.	woolly.	—	—	G. tr.
8. juniperinum.	juniper leaved.	—	—	G. tr.
FABRICIA				
1. laevigata.	smooth.	New Holland.	—	G. tr.
METROSIDEROS				
1. hispida.	rough.	New Holland.	July, August.	G. tr.

ICOSANDRIA MONOGYNIA.

METROSIDEROS				
2. linearis.	linear leaved.	New Holland.	July, August.	*G. tr.*
3. lanceolata.	spear leaved.		March, June.	*G. tr.*
4. saligna.	willow leaved.			*G. tr.*
5. diffusa.	spreading.	New Zealand.		*G. tr.*
6. angustifolia.	narrow leaved.	C. G. H.		*G. tr.*
7. viminalis.	twiggy.	New Holland.		*G. tr.*
8. capitata.	headed.			*G. tr.*
PSIDIUM GUAVA				
1. pyriferum.	pear fruited.	Indies.	June.	*S. tr.*
2. pomiferum.	apple fruited.			*S. tr.*
EUGENIA				
1. Jambos.	willow leaved.	E. India.	May, June.	*G. tr.*
2. uniflora.	one flowered.	Brazile.		*S. tr.*
MYRTUS				
1. communis.	common.	S. Europe.	July, August.	*G. tr.*
var. *Italic.*	*Italian ditto.*			*G. tr.*
—— *Lusitan.*	*Portugal ditto.*			
—— *mucron.*	*rosemary leaved ditto.*			
—— *Boetic.*	*orange leaved ditto.*			
—— *Belgic.*	*broad leaved Dutch ditto.*			
2. tomentosa.	woolly.	China. Cochinchina.	June, July.	*G. tr.*
3. coriacea.	leathery leaved.	Hispaniolia, St. Lucia.		*S. tr.*

ICOSANDRIA MONOGYNIA.

MYRTUS				
4. Pimenta.	Jam. pepper, or *all spice*.	W. India.	July.	*S. tr.*
5. androfœmoides.	Tutfan leaved.	Ceylon.		*S. tr.*
CALYPTRANTHUS				
1. Suzygium.	fhrubby Suzygium.	Jamaica.		*S. tr.*
2. Chytraculia.	baftard green heart.			*S. tr.*
EUCALYPTUS				
1. piperata.	pepper mint.	New Holland.	July.	*G. tr.*
2. obliqua.	oblique leaved.			*G. tr.*
3. haemaftoma.	leathery leaved.			*G. tr.*
PUNICA				
1. nana.	dwarf.	Caribee Iflands.	July, Sept.	*G. tr.*

ICOSANDRIA PENTAGYNIA.

MESPILUS, or *Medlar*.				
1. Japonica.	Japan.	Japan.		*G. tr.*
TETRAGONIA				
1. expanfa.	horned.	{ New Zealand, Friendly Iflands, Japan. }	Auguft, May.	*G. bi.*
MESEMBRYANTHEMUM FIG-MARYGOLD				
1. linguaeforme.	tongue leaved.	C. G. H.	M. P. S.	*G. pe.*
2. latum.	broad leaved.			*G. pe.*

ICOSANDRIA PENTAGYNIA.

MESEMBRYANTHEMUM. FIG-MARYGOLD

No.	Species	Common name	Locality	Flowering	Code
3.	rostratum.	heron-beaked.	—	Sep.	G. pe.
4.	caninum.	dog-chap.		August, October.	G. pe.
5.	tigrinum.	tiger-chap.		Sept.	G. pe.
6.	murinum.	mouse-chap.			G. pe.
7.	albidum.	white.		July, August.	G. pe.
8.	bellidiflorum.	daisy-flowered.		May, August.	G. pe.
9.	Tripolium.	plain leaved.		June, Sept.	G. bi.
10.	calamiforme.	quill-leaved.		May, August.	G. tr.
11.	minimum.	leaf.		June, July.	G. tr.
12.	pinnatifidum.	jagged leaved.	Greece, near Athens.	July, August.	G. an.
13.	cryftallinum.	ice plant.	C. G. H.	June, Sept.	G. an.
14.	glabrum.	smooth.		July.	G. an.
15.	expanfum.	houfe-leek leaved.		June, Aug.	G. pa.
16.	geniculiflorum.	joint-flowering.			G. tr.
17.	umbellatum.	umbelled.		June, Sept.	G. tr.
18.	bicolorum.	two-coloured.		May, Sept.	G. tr.
19.	tuberofum.	tuberous rooted.		June, October.	G. tr.
20.	tenuifolium.	flender leaved.		June, Sept.	G. tr.
21.	ftipulaccum.	upright fhrubby.			G. tr.
22.	loreum.	leathery ftalked.		Auguft.	G. pe.
23.	veruculatum.	fplit leaved.		May, July.	G. tr.

ICOSANDRIA PENTAGYNIA.

MESEMBRYANTHEMUM. FIG-MARYGOLD.

No.	Name	Description	Locality	Flowering	
24.	echinatum.	echinated.	C. G. H.	July, October.	G. tr.
25.	viridiflorum.	green flowered.			G. tr.
26.	splendens.	shining.		June, Aug.	G. tr.
27.	micans.	glittering.		———	G. tr.
28.	grossum.	gouty.			G. tr.
29.	barbatum.	shrubby bearded.		M P. S.	G. tr.
30.	falcatum.	sickle leaved.		June, August.	G. tr.
31.	glomeratum.	glomerated.			G. tr.
32.	reptans,	creeping.		———	G. tr.
33.	deflexum.	bending.		July, Oct.	G. tr.
34.	spinosum.	thorny.		June, Sept.	G. tr.
35.	glaucum.	glaucous leaved.		June, July.	G. tr.
36.	spectabile.	large purple flowered.		July, Oct.	G. tr.
37.	emarginatum.	notched flowered.		June, July.	G. tr.
38.	aureum.	golden.		June, August.	G. tr.
39	serratum.	serrated.			G. tr.
40.	scabrum.	rough leaved.		June, August.	G. tr.
41.	uncinatum.	small hooked.		June, Sept.	G. tr.
42.	pugioniforme.	dagger leaved.	New Holland.	May, Aug.	G. tr.
43.	aequilaterale.	long green leaved.	C. G. H.		G. tr.
44.	filamentosum.	thready.		July, Sept.	G. tr.

D d

ICOSANDRIA PENTAGYNIA.

MESEMBRYANTHEMUM, FIG-MARYGOLD

			C. G. H.		
45.	deltoides.	delta leaved.	.	—	May, Aug.	G. tr.
46.	acinaciforme.	fimitar leaved.			Sept.	G. tr.
47.	fubulatum.	oval leaved.		—	May, Aug.	G. tr.
48.	gibbofum.	gibbous.		—		G. tr.
49.	fcalpatum.	broadeft tongue,	.	—	M. P. S.	G. tr.
50.	cornutum.	horned.				
51.	tricolorum.	three coloured.	.	—		G. tr.
52.	fpeciofum.	handfome.				
53.	perfoliatum.	great hooked.	.	—	June, Sept.	G. tr.
54.	tenellum.	fmall connate.	'	—	July.	G. tr.
55.	compreffum.	compreffed.		—	July, Oct.	G. tr.
56.	mutabilis.	changeable.		—	July, Sept.	G. tr.
57.	lacerum.	purple ferrated.	.	—	Aug. Sept.	G. tr.
58.	maximum.	great.	.	—	Sept.	G. tr.
59.	cylindricum.	cylindric.	.	—		G. tr.
60.	coccineum.	fcarlet flowered.	'	—	Jun. Oct.	G. tr.
61.	fubulatoides.					
62.	reflexum.	fhort leaved horned.		—	Mar. May.	G. tr.
63.	maculatum.	fpotted.	'			
64.	monoliforme.	necklace.		—		
65.	hirfutum.	fmall bearded.	'	—	June, Oct.	G. tr.

ICOSANDRIA POLYGYNIA.

POLYGYNIA.

ROSÆ				
1. lucida.	shining leaved.	- N. America.	June, July.	C. tr.
2. semperflorens.	red China.	- China.	March, Nov.	G. tr.
3. Chinensis.	pale China.	- China.	———	G. tr.
4. Indica.	Indian.	- China.	———	G. tr.
CALYCANTHUS				
1. præcox.	early flowering.	- Japan.	Jan. Mar.	G. tr.

CLASSIS XIII.

POLYANDRIA MONOGYNIA.

CAPPARIS				
1. spinosa.	prickly.	- S. of Europe.	May, June.	G. tr.
2. Cynophallophora.	laurel leaved.	- W. India.		S. tr.
3. linearis.	linear.	- Carthagena.		S. tr.
CALOPHYLLUM				
1. Calaba.	bay leaved.	- Indies.		S. tr.

POLYANDRIA MONOGYNIA.

BIXA	arnotta.	S. America.	-	S. tr.
1. Orellana.				
GREWIA				
1. Occidentalis.	elm leaved.	C. G. H.	June, August.	G. tr.
2. Orientalis.	Oriental.	E. India.		S. tr.
LARGERSTROEMIA				
1. Indica.	Indian.	China, Cochinchina, Japan.	August, Sept.	S. tr.
2. reginae.	oblong leaved.	Calcutta, Java.		S. tr.
THEA				
1. Bohea.	Bohea.	Japan, China.	Mar. May.	G. tr.
2. viridis.	green.	China.	Aug. Sept.	G. tr.
CISTUS, or *Rock-rose*				
1. villosus.	villose.	Italy, Spain, Africa.	July,	G. tr.
2. vaginatus.	sheathed.	Teneriffe.	April, June.	G. tr.
3. libanotis.	rosemary leaved.	Spain.	June.	G. tr.
4. Syriacus.	Syrian.			G. tr.
CORCHORUS				
1. olitorius.	briefly leaved.	Asia, Africa, America.	June, Aug.	S. am.
2. capsularis.	heart-leaved.	India.	June, July.	S. am.

POLYANDRIA TETRAGYNIA.

TETRACUA				
1. farmentofa.	creeping rooted.	- Ceylon.		S. tr.

POLYGYNIA.

DILLENIA				
1. fcandens.	climbing.	- New Holland.	June, July.	G. tr.
2. fpeciofa.	handfome.	- Malabar, Java.		G. tr.
ILICIUM				
1. anifatum.	anifeed.	- Japan, China.		S. tr.
2. floridanum.	red flowered.	- Florida.	April, Auguft.	G. tr.
MAGNOLIA				
1. obovata.	purple.	- Japan, China.	Apr. May.	G. tr.
2. Plumieri.	Plumier's.	- St. Lucia, Martinique, Guadaloupe.		S. tr.
ANNONA				
1. fquamofa.	undulated.	- S. America.		S. tr.
2. hexapetala.	long leaved.	- E. India, China.		S. tr.
3. muricata.	prickly.	- S. America.		S. tr.
ATRAGENE				
1. Capenfis.	Cape.	• C. G. H.	Mar. Apr.	G. tr.

POLYANDRIA POLYGYNIA.

ATRAGENE				
2. Zeylanica.	Ceylon.	Ceylon	-	G. tr.
CLEMATIS	VIRGIN'S BOWER			
1. florida.	large flowering.	Japan.	M. P. S.	G. tr.
2. calycina.	Minorca.	Minorca.		G. tr.
ADONIS				
1. veficatoria.	blifter.	C. G. H.	Mar. Apr.	G. pt.

CLASSIS XIV.

DIDYNAMIA GYMNOSPERMIA.

TEUCRIUM	GERMANDER			
1. fruticans.	fhrubby.	Spain, Sicily.	June, Sept.	G. tr.
2. latifolium.	broad leaved tree.	Spain.	July, Sept.	G. tr.
3. Marum.	Marum.	———		G. tr.
v. fol. varieg.	variegated ditto.	———		
4. Afiaticum.	Afiatic.	Madeira.	June, Oct.	G. tr.
5. betonicum.	betony leaved.		May, Auguft	G. tr.
6. abutiloides.	mulberry leaved.		April, May.	C. tr.
7. Maffilienfe.	fweet fcented.	S. of France.	June, July.	G. tr.
8. flavum.	yellow flowered.	S. of Europe.	July, Sept.	G. tr.

DIDYNAMIA GYMNOSPERMIA.

			M. P. S.	
TEUCRIUM	GERMANDER			
9. Polium.	Poley.	Levant, S. of Europe.	July, Sept.	*G. tr.*
WESTRINGIA	WESTRINGIA			
1. rosmarinifolia.	rosemary leaved.	New Holland.	*M. P. S.*	*G. tr.*
LAVANDULA	LAVENDER			
1. Stoechas.	French.	S. of Europe.	May, June.	*G. tr.*
2. dentata.	tooth leaved.	Spain, Levant.	June, Sept.	*G. tr.*
3. multifida.		Spain, Canary Islands.	July, Nov.	*G. tr.*
SIDERITIS	IRON WORT			
1. Canariensis.	Canary.	Madeira, Canary Islands.	May, August.	*G. tr.*
2. candicans.	mullein leaved.	Madeira.	April, July.	*G. tr.*
BYSTROPOGON	BYSTROPOGON			
1. punctatum.	cluster flowered.	Madeira.	July, Sept.	*G. tr.*
2. Canariense.	Canary.	Madeira, Canary Islands.	June, August.	*G. tr.*
PHLOMIS	PHLOMIS			
1. purpurea.	purple.	Spain, Italy.	June, August.	*G. tr.*
2. leonorus.	lion's tail.	C. G. H.	Oct. Dec.	*G. tr.*
3. Leonites.	dwarf shrubby.		June, July.	*G. tr.*
ORIGANUM	MARJORAM			
1. Ægyptiacum.	Egyptian.	Egypt.	June, August.	*G. p.*
v. fol. varieg.	variegated ditto.			
2. Dictamnus.	dittany of Candia.	Island of Candia.	June, July.	*G. tr.*
3. Marjorana.	sweet, or knotted.			*G. tr.*

DIDYNAMIA GYMNOSPERMIA.

ORIGANFM	MARJORAM			
4. hybridum.	bastard.	-	June, Sept.	G. pe.
THYMUS	THYME			
1. Mastichina.	mastick.	- Spain.	July, Sept.	G. tr.
DRACOCEPHALUM	DRAGON's HEAD			
1. Canariense.	balm of Gilead.	Canary Islands.	July, Sept.	G. tr.
PLECTRANTHUS	PLECTRANTHUS			
1. fruticosus.	shrubby.	- C. G. H.	June, Sept.	G. tr.
PRASIUM	PRASIUM			
1. majus.	great Spanish hedge nettle.	Spain, Italy.	June, Aug.	G. tr.
2. minor.	small ditto.	-		G. tr.
OCYMUM	BASIL			
1. gratissimum.	shrubby.	- E. India.	July, August.	S. tr.
2. Basilicum.	common sweet.	- India, Persia.		G. an.
3. minimum.	bush.	- E. India.		S. an.
4. sanctum.	sacred herb.	-	Sept.	S. an.

ANGIOSPERMIA.

CHELONE	CHELONE			
1. Ruellioides.	Ruellus's.	Chili.	June, August.	G. pe.
2. campanuloides.	bell flowered.	Mexico.	June, August.	G. tr.

DIDYNAMIA ANGIOSPERMIA.

ANTIRRHINUM	TOAD FLAX			
1. macrocarpum.	large fruited.	C. G. H.	Mar.	G. pe.
2. Afarina.	heart leaved.	Italy.	July.	G. pe.
CITHAREXYLUM	FIDDLE WOOD			
1. quadrangulare.	square ftalked.	Jamaica.		S. tr.
CYRILLA	CYRILLA.			
1. pulchella.	scarlet flowered.	Jamaica.	July, Aug.	S. pe.
GLOXINIA	GLOXINIA			
1. maculata.	spotted.	S. America.	July, August.	S. pe.
MAURANDYA	MAURANDYA			
1. femperflorens.	climbing.	Mexico.	May, Sept.	G. tr.
BRUNFELSIA	BRUNFELSIA			
1. Americana.	American.	W. India.	June, July.	S. tr.
CELSIA	CELSIA			
1. Arcturus.	scollop leaved.	Crete.	July, Sept.	G. bi.
2. linearis.	linear leaved.	Peru.	M. P. S.	G. tr.
3. Orientalis.	cut leaved.	Levant.	July, August.	G. an.
4. urticæfolia.	nettle leaved.	Peru.	M. P. S.	G. tr.
DIGITALIS	FOX-GLOVE			
1. obscura.	willow leaved.	Spain.	July, Aug.	G. tr.
2. sceptrum.	shrubby.	Madeira.	July, Aug.	G. tr.
3. Canariensis.	Canary.	Canary Islands.	June, July.	G. tr.
CLERODENDRUM	CLERODENDRUM			
1. infortunatum.		E. India.		S. tr.

E c

DIDYNAMIA ANGIOSPERMIA.

CLERODENDRUM	CLERODENDRUM			
2. dichotomum.		Japan.		S. tr.
BIGNONIA	TRUMPET FLOWER			
1. fempervirens.	yellow fweet fcented.	N. America.	June, July.	G. tr.
2. pandorana.	clufter flowered.	Norfolk ifland.	Mar. June.	G. tr.
3. leucoxylon.	digiated.	W. India.	July, Aug.	S. tr.
SCROPHULARIA	FIG-WORT			
1. fambucifolia.	elder leaved.	Spain, Portugal.	July, Sept.	G. pe.
HALERIA	HALERIA			
1. lucida.	fhining leaved.	C. G H.	June, Aug.	G. tr.
LANTANA	LANTANA			
1. Camara.	various coloured.	W. India.	June, Sept.	G. tr.
2. involucrata.	round leaved.		May, July.	S. tr.
3. aculeata.	prickly.		April, Nov.	S. tr.
v. fl. alb.	white flowered ditto.			
4. Africana.	ilex leaved.	C. G. H.	Feb. Nov.	G. tr.
CAPRARIA	CAPRARIA			
1. lanceolata.	willow leaved.	C. G. H.		G. tr.
2. undulata.	wave leaved.		March, July.	G. tr.
SELAGO	SELAGO			
1. corymbofa.	heath leaved.	C. G. H.	June, July.	G. tr.
MANULEA	MANULEA			
1. tomentofa.	woolly.	C. G. H.	May, Nov.	G. tr.

DIDYNAMIA ANGIOSPERMIA.

HEBENSTRETIA	HEBENSTRETIA				
1. dentata.	tooth leaved.	-	C. G. H.	Mar. Nov.	G. bi.
BROWALLIA	BROWALLIA				
1. elata.	upright.	-	S. America.	July, August.	S. an.
2. demiffa.	spreading.	-	———		S. an.
MIMULUS	MONKEY FLOWER				
1. aurantiacus.	orange.	-	Peru.	M. P. S.	G. tr.
BARLERIA	BARLERIA				
1. Prionitis.	prickly.	-	E. India.	August.	S. tr.
DURANTA	DURANTA				
1. Plumieri.	smooth.	-	S. America.	Oct.	S. tr.
2. Ellifia.	prickly.	-	W. India.	August.	S. tr.
VOLKAMERIA	VOLKAMERIA				
1. aculeata.	prickly.	-	W. India.	August, Oct.	S. tr.
2. inermis.	smooth.	-	E. India.	July, Aug.	S. tr.
3. Kaempferiana.	scarlet flowered.	-	———		S. tr.
VITEX	CHASTE-TREE				
1. trifolia.	three leaved.	-	E. India.		S. tr.
BONTIA	WILD OLIVE				
1. daphnoides.	Barbadoes.	-	W. India.	June.	S. tr.
COLUMNEA	COLUMNEA				
1. hirfuta.	hairy.	-	Jamaica.	Aug. Nov.	S. tr.
MELIANTHUS	HONEY-FLOWER.				
1. major.	great.	-	C. G. H.	May, June.	G. tr.

Ee 2

DIDYNAMIA ANGIOSPERMIA.

MELIANTHUS	HONEY-FLOWER				
2. minor.	fmall.	-	C. G. H.	Auguft.	G. tr.
THUNBERGIA					
1. fragrans.	fweet fmelling.	-	Coaft of Coromandel.		S. tr.

CLASSIS XV.

TETRADYNAMIA SILICULOSA.

LEPIDIUM	PEPPER-WORT				
1. fubulatum.	awl-leaved.	-	Spain.	July, Auguft.	G. tr.
IBERIS	CANDY-TUFT				
1. femperflorens.	fhrubby.	-	Perfia, Sicily.	M. P. S.	G. tr.
VELLA					
1. Pfeudo-cytifus.	fhrubby.	-	Spain.	April, May.	G. tr.

SILIQUOSA.

CHEIRANTHUS	STOCK				
1. mutabilis.	changeable flower'd.	-	Madeira.	March, May.	G. tr.
2. feneftralis.	clufter leaved.	-		July, Aug.	G. bi.

TETRADYNAMIA SILIQUOSA.

CHEIRANTHUS	STOCK			
3. triftis.	dark flowered.	- S. of Europe.	May, July.	*G. bi.*
CRAMBE	COLE-WORT			
1. fruticofa.	fhrubby.	- Madeira.	*M. P. Y.*	*G. tr.*
CLEOME	CLEOME			
1. vifcofa.	clammy.	- Ceylon.	June, July.	*S. an.*
2. gigantea.	gigantic.	- S. America.	——	*S. tr.*
3. fpinofa.	prickly.	- W. India.	——	*S. bi.*
4. pentaphylla.	five leaved.	- Indies.	——	*S. an.*

CLASSIS XVI.

MONADELPHIA PENTANDRIA.

HERMANNIA	HERMANNIA			
1. micans.	gliftening.	- C. G. H.	May, Auguft.	*G. tr.*
2. candicans.	hoary.	——	April, June.	*G. tr.*
3. althaeifolia.	althaea leaved.	• ——	Auguft, Oct.	*G. tr.*
4. plicata.	plaited.	: ——		*G. tr.*
5. hyffopifolia.	hyffop leaved.	: ——	April, May.	*G. tr.*
6. alnifolia.	alder leaved.	• ——	March, May.	*G. tr.*

MONADELPHIA PENTANDRIA.

HERMANNIA				
7. lavandulifolia.	lavender leaved.	- C. G. H.	May, Sept.	G. tr.
8. decumbens.	decumbent.	- ———		G. tr.
9. odorata.	sweet scented.	- ———	May, August.	G. tr.
10. trifurcata.	three forked.	- ———		
11. rotundifolia.	round leaved.	- ———		
12. angularis.	angular.	- ———	May, July.	G. tr.
13. denutata.	smooth.	- ———		
ERODIUM	**HERON'S BILL**			
1. crassifolium.	thick leaved.	- Island of Cyprus.	May, June.	G. pe.
2. incarnatum.	flesh coloured.	- C. G. H.	July, August.	G. pe.
3. trilobatum.	three lobed.	- ———		

HEPTANDRIA.

PELARGONIUM	**STORK'S BILL**			
1. lobatum.	lobed.	- C. G. H.	July, August.	G. pe.
2. triste.	night smelling.	- ———		G. pe.
3. roseum.	rose.	- ———	April, June.	G. pe.
4. punctatum.	dotted.	- ———	April, May.	G. pe.
5. odoratissimum.	sweet scented.	- ———	May, Oct.	G. pe.
6. grossularioides.	gooseberry leaved.	- ———		G. pe.

MONADELPHIA HEPTANDRIA.

		STORK'S BILL	C. G. H.		
	PELARGONIUM				
7.	anceps.	sharp edged.	—	May, June.	G. tr.
8.	coriandrifolium.	coriander leaved.	—	April, Sept.	G. pa.
9.	myrrhifolium,	myrrh-leaved.		May, August.	G. tr.
10.	betonicum.	betony leaved.		June, August.	G. bi.
11.	tenuifolium.	fine leaved.	—	May, August.	G. tr.
12.	apiifolium.	parsley.	—		G. tr.
13.	carnosum.	fleshy stalked.	—	June, August.	G. tr.
14.	gibbosum.	gouty.	—	Mar. June.	G. tr.
15.	fulgidum.	celandine leaved.	—		G. tr.
16.	papilonaceum.	butterfly.	—	March, July.	G. tr.
17.	monstrum.	clustered.	—		G. tr.
18.	tricolor.	three coloured.	—	June, August.	G. tr.
19.	vitifolium.	vine leaved.	—	April.	G. tr.
20.	capitatum.	rose scented.	—		G. tr.
21.	glutinofum.	clammy.	—	May, June.	G. tr.
22.	cucullatum.	hooded.		May, Sept.	G. tr.
23.	angulosum.	marsh-mallow.	—	July, August.	G. tr.
24.	echinatum.	prickly stalked.	—	May, August.	G. tr.
25.	tetragonum.	square stalked.	—	June, August.	G. tr.
26.	peltatum.	peltated.	—		G. tr.
	v. fol. varieg.	variegated ditto.	—		G. tr.

MONADELPHIA HEPTANDRIA.

PELARGONIUM	STORK'S BILL	C. G. H.		
27. lateripes.	ivy leaved.	.	July, August.	G. tr.
28. grandiflorum.	large flowered.	.		G. tr.
29. hepaticifolium.	hepatica leaved.			June, July.	G. tr.
30. cortusæfolium.	cortusa leaved.				G. tr.
31. crassicaule.	thick stalked.				G. tr.
32. betulinum.	birch leaved.			May, July.	G. tr.
33. tomentosum.	woolly.			June, August.	G. tr.
34. tricuspidatum.	three pointed.			May, August.	G. tr.
35. scabrum.	rough.			August, Sept.	G. tr.
36. crispum.	curled			July, Nov.	G. tr.
37. exstipulatum.	soft leaved trifid.			May, Aug.	G. tr.
38. incisum.	cut leaved.			M. P. S.	G. tr.
39. fragile.	brittle stalked.			July, Sept.	G. tr.
40. ciliatum.	ciliated.				G. tr.
41. radula.	rasp.			Mar. Aug.	G. tr.
42. alchimilloides.	lady's mantle.			May, Oct.	G. tr.
43. acetosum.	sorrel.			May, Sept.	G. tr.
44. bicolor.	two coloured.			July, Aug.	G. tr.
45. inquinans.	scarlet.			May, Sept.	G. tr.
46. adulterinum.	villous.			April, May.	G. tr.
47. flavum.	carrot leaved.			July, Sept.	G. pe.

MONADELPHIA HEPTANDRIA.

PELARGONIUM	STORK'S BILL			
48. denticulatum.	toothed.	-	C. G. H.	June, July.	G. tr.
49. ternatum.	ternate.	-	——	June, July.	G. tr.
50. Zonale.	horse-shoe.	-	——	May, Nov.	G. tr.
51. cordifolium.	heart leaved.	-	——	Mar. June.	G. tr.
52. hybridum.*	bastard.	-	——	May, Sept.	G. tr.

OCTANDRIA.

AITONIA				
1. Capensis.	cluster leaved.	-	C. G. H.	M. P. S.	G. tr.

DECANDRIA.

GERANIUM	CRANE'S BILL			
1. anemonifolium.	anemone leaved.	-	Madeira.	May, Sept.	G. α.

* Besides many *varieties* of the *species* of Erodium and Pelargonium.

F f

MONADELPHIA DODECANDRIA.

PENTAPETES
1. Phoenicea.   scarlet-flowered. - E. India.   June, July.   *S. bi.*

PTEROSPERMUM
1. acerifolium.   maple leaved. - E. India.   *S. tr.*

POLYANDRIA.

BOMBAX					
Ceiba.	SILK COTTON TREE five leaved.	-	Indies.		*S. tr.*
SIDA					
1. spinosa.	prickle seeded.	-	East India.	July, Sept.	*S. an.*
2. carpinifolia.	horn beam leaved.	-	———		*S. bi.*
3. Abutilon."	broad leaved.	-	———		*S. an.*
4. cristata.	crested.	-	Mexico.		*S. an.*
5. triquetra.	triangular stalked.	-	W. India.	July, August.	*S. bi.*
6. cordifolia.	heart leaved.	-	E. India.	July, Sept.	*S. an.*
7. Indica.	Indian.	-	———	June, Aug.	*S. an.*
MALACHRA					
capitata.	headed.	-	W. India.	Aug. Sept.	*S. an.*
MALVA					
1. spicata.	spiked.	-	Jamaica.	July, August.	*S. bi*
2. Capensis.	Cape.	-	C. G. H.	*M. P. S.*	*G. tr.*

MONADELPHIA POLYANDRIA.

		C. G. H.	M. P. S.	
MALVA				
3. grossulariæfolia.	gooseberry leaved.			G. tr.
4. bryonifolia.	bryony leaved.			G. tr.
LAVATERA				
1. olbia.	downy leaved.	S. of Europe.	June, Oct.	G. tr.
URENA				
1. lobata.	angular leaved.	China, E. India.	July, Aug.	S. tr.
GOSSYPIUM				
1. arboreum.	tree.	E. India.		S. tr.
2. Barbadense.	Barbadoes.	Barbadoes.	Sept.	S. bi.
HIBISCUS				
1. praemorsus.	Cape.	C. G. H.	June. Aug.	G tr.
2. mutabilis.	changeable.	E. India.	Nov. Dec.	S. tr.
3. spinifex.	prickly seeded.	W. India.	July.	S. tr.
4. Solandra.	maple leaved.	Bourbon.	August.	S au.
5. ficulneus.	fig leaved.	Ceylon.	June, July.	S. tr.
6. speciosus.	scarlet.	S. Carolina.	June, July.	G. pe.
7. Manihot.	palmated.	China, Japan.	August.	£. tr.
8. esculentus.	esculent.	W. India.	June, July.	G. an.
9. cannabinus.	hemp leaved.	E. India.	July. Aug.	S. an.
10. Rosa Sinensis.	China rose.			S. tr.
var. fl. plen.	double flowered ditto.			S. tr.

MONADELPHIA POLYANDRIA

ACHANIA				
1. mollis.	foft leaved. -	W. India.	Aug. Sept.	S. tr.
2. Malvavifcus.	fcarlet flowering. -	Jamaica.	M. P. S.	S. tr.
CAMELLIA	JAPAN ROSE			
1. Japonica.	common. -	China.	Feb. May.	G. tr.
var. fl. alb. plen.	double flowered white ditto. —			

CLASSIS XVII.

DIADELPHIA HEXANDRIA.

FUMARIA				
1. veficaria.	bladdered. -	C. G. H.	June, July.	G. am.

OCTANDRIA

POLYGALA	MILK-WORT			
1. myrtifolia.	myrtle leaved. -	C. G. H.	May, Oct.	G. tr.
2. fpinofa.	prickly. -		Feb. March.	G. sr.

DIADELPHIA DECANDRIA.

DECANDRIA.

ABRUS	WILD LIQUORICE			
1. precatorius.	Jamaica.	- Indies.	Auguft, Sept.	*S. tr.*
ERYTHRINA	CORAL TREE			
1. herbacea.	American.	- S. Carolina.	May, June.	*S. tr.*
2. Corallodendrum.	fmooth leaved.	- W. India.		*S. tr.*
SPARTIUM	BROOM			
1. monofpermum.	fingle feeded.	- Spain, Portugal.	June, July.	*G. tr.*
2. fpinofum.	prickly.	- S. of Europe.	June, July.	*G. tr.*
5. virgatum.	twiggy.	- Madeira.	March, June.	*G. tr.*
GENISTA	GENISTA			
1. Canarienfis.	Canary.	- Canary Iflands.	May, June.	*G. tr.*
2. linifolia.	flax leaved.	- Spain.	April, July.	*G. tr.*
ASPALATHUS	ASPALATHUS			
1. mucronata.				
CROTALLARIA	CROTALLARIA			
1. fagittalis.	arrow.	- Jamaica, S. America.	June, July.	*S. an.*
2. ilicifolia.	ilex leaved.	- E. India.	July.	*S. an.*
3. biflora.	two flowered.	-		*S. an.*
ONONIS	REST-HARROW			
1. Natrix.	yellow flowered.	- S. of Europe.	May. Sept.	*G. tr.*
2. rotundifolia.	round leaved.	- Switzerland.	May, June.	*G. tr.*

DIADELPHIA DECANDRIA.

ANTHYLLIS	**KIDNEY VETCH**			
1. Erinacea.	hedge-hog.	Spain.	April, May.	G. tr.
2. Barba-jovis.	Jupiter's beard.	Levant.		G. tr.
3. Cytisoides.	cytisus leaved.	Spain.	April, June.	G. tr.
4. Hermaniæ.	lavender leaved.	Levant.		G. tr.
DOLICHOS	**DOLICHOS**			
1. Lablab.	black feeded.	Egypt.	June, July.	G. an.
GLYCINE	**GLYCINE**			
1. bituminosa.	clammy.	C. G. H.	April, Sept.	G. tr.
2. bimaculata.	two spotted.		Mar. May.	G. tr.
3. rubicunda.	red.	New Holland.	March, July.	G. tr.
4. coccinea.	scarlet.		March, May.	G. tr.
PLATILOBIUM	**FLAT PEA**			
1. formosum.	large flowered.	New Holland.	June, July.	G. tr.
GEOFFROYA	**BASTARD CABBAGE TREE**			
1. inermis.	smooth.	Jamaica.		S. tr.
LIPARIA	**LIPARIA**			
1. villosa.	villous.	C. G. H.	June, July.	G. tr.
CYTISUS	**CYTISUS**			
1. foliosus.	leafy.	Canary Islands.	July, August.	G. tr.
2. divaricatus.	clammy podded.	Spain, Madeira.		G. tr.
3. Cajan.	pidgeon pea.	Indies.		S. an.
4. proliferus.	silky.	Canary Islands.	April, May.	G. tr

DIADELPHIA DECANDRIA.

COLITEA	BLADDER-SENNA			
1. frutescens.	scarlet flowered. •	C. G. H.	June, July.	G. tr.
2. perennans.	perennial.	Africa.	August.	G. tr.
ROBINIA	ROBINIA, or *false Acacia*			
1. violacea.	ash leaved. •	W. India.		S. tr.
CORONILLA	CORONILLA			
1. valentina.	nine leaved. •	Spain, Italy.	Feb. Mar.	G. tr.
2. glauca.	seven leaved. •	S. of France.	May, Oct.	G. tr.
HIPPOCREPIS	HORSE-SHOE-VETCH			
1. Balearica.	shrubby. •	Minorca.	May, June.	G. tr.
ÆSCHYNOMENE	ÆSCHYNOMENE			
1. grandiflora.	great flowered. •	E. India.		S. bi.
2. Sesban.	Egyptian. •	Egypt.	July, August.	S. tr.
3. Indica.	Indian. •	Indies.		S. tr.
HEDYSARUM	HEDYSARUM			
1. vespertilionis.	bat-winged. •	Cochinchina.	July.	S. bi.
2. gyrans.	moving. •	E. India.	July, Aug.	S. bi.
3. gangeticum.	oval leaved. •			S. bi.
INDIGOFERA	INDIGO			
1. psoraloides.	long spiked. •	C. G. H.	April, July.	G. tr.
2. amoena.	scarlet flowered. •		April, June.	G. tr.
3. Australis.	Botany bay. •	New Holland.	April, May.	G. tr.
4. sarmentosa.	dwarf. •	C. G. H.	June.	G. pe.

DIADELPHIA DECANDRIA.

GALEGA	**GOAT'S RUE.**			
1. grandiflora.	great flowered.	C. G. H.	May, Sept.	G. bi.
2. villofa.	villous.	Indies.		G. tr.
PSORALEA	**PSORALEA**			
1. bracteata.	oval spiked.	C. G. H.	June, July.	G. tr.
2. bituminofa.	bituminous.	Italy, France.	M. P. S.	G. tr.
3. glandulofa.	striped flowered.	Peru.	May, Auguft.	G. tr.
4. corylifolia.	hazle leaved.	E. India.	June, July.	S. bi.
5. decumbens.	trailing.	C. G. H.	May, June.	G. tr.
6. pinnata.	wing leaved.	——	May, July.	G. tr.
7. angustifolia.	narrow leaved.	——	May, Auguft.	G. tr.
LOTUS	**BIRD'S-FOOT TREFOIL,**			
1. Creticus.	filver leaved.	Spain, Levant.	June, Sept.	G. tr.
2. jacobaeus.	dark flowered.	Cape Verd Iflands.	M. P. S.	G. tr.
3. hirfutus.	hairy.	S. of Europe.	June, Auguft.	G. tr.
4. Dorycnium.	fhrubby.	——	July, Sept.	G. tr.
MEDICAGO	**MEDICK**			
1. arborea.	tree.	Italy.	May, Nov.	G. tr.
2. marina.	fea.	Mediter. coafts.	June, July.	G. pe.

CLASSIS XVIII.

POLYADELPHIA PENTANDRIA.

THEOBROMA baftard cedar.				
1. Guazuma.	-	Indies.		S. tr.

DODECANDRIA.

MONSONIA fine leaved.		C. G. H.	March, May.	G. h.
1. fpeciofa.	:			G. h.
2. fiha. hairy.	:	———		G. h.
3. lobata, lobed.	:	———		

ICOSANDRIA.

CITRON TREE fhaddock.		E. India.	May, July.	G. tr.
1. decumana.	-			
2. trifoliata. three leaved.	:	Japan.		S. tr.

G g

POLYADELPHIA POLYANDRIA.

POLYANDRIA.

MELALEUCA

1. ericifolia.	heath leaved.	New Holland.		G. tr
2. linarifolia.	toad flax leaved.	——		G. tr.
3. hypericifolia.	St. John's wort leaved.	——	July, August.	G. tr.
4. squarrosa.	scaley.	——		G. tr.

HYPERICUM ST. JOHN'S WORT

1. Balearicum.	wart leaved.	Majorca.	Mar. Sept.	G. tr.
2. coris.	heath leaved.	S. of Europe.	M. P. S.	G. tr.
3. monogynum.	Chinese.	China.	Mar. Sept.	G. tr.
4. Canariense.	Canary.	Canary Islands.	July, Aug.	G. tr.
5. tomentosum.	woolly.	S. of Europe.	July, Sept.	G. pe.
6. reflexum.	reflex leaved.	Teneriffe.	M. P. S.	G. tr.

ASCYRUM ST. ANDREW'S CROSS

1. Crux-andreae.	common.	N. America.	July, August.	G. tr.

CLASSIS XIX.

SYNGENESIA POLYGAMIA ÆQUALIS.

SONCHUS SOW THISTLE

1. fruticosus.	shrubby.	Madeira.	April, July.	G. tr.

SYNGENESIA POLYGAMIA ÆQUALIS.

CREPIS	CREPIS			
1. filiformis.	fine leaved.	•	Madeira.	G. tr.
CICHORIUM	ENDIVE			
1. spinosum	prickly.	-	Sicily, Crete.	G. tr.
CARDUUS	THISTLE			
1. casobonae.	fish.	•	S. of Europe.	G. bi.
CARTHAMUS	CARTHAMUS			
1. salicifolius.	willow leaved.	•	Madeira.	G. tr.
SPILANTHUS	SPILANTHUS			
1. Pseudo-Acmella.	spear leaved.	•	Ceylon.	S. an.
2. oleracea.	eatable.	•	E. India.	S. an.
CACALIA	CACALIA			
1. articulata.	jointed stalked.	•	C. G. H.	G. tr.
2. Anteuphorbium.	oval leaved.	•	Canary Islands.	S. tr.
3. Kleinia.	cabbage tree.	•	C. G. H.	S. tr.
4. Ficoides.	fig marygold.	•		G. tr.
5. repens.	glaucous.	•		G. tr.
STÆHELINA	STÆHELINA			
1. Chamaepeuce.		'	Crete.	G. tr.
PTERONIA	PTERONIA			
1. camphorata.	camphire smelling.	-	C. G. H.	G. tr.
2. oppositifolia.	forked.	-		G. tr.

SYNGENESIA POLYGAMIA ÆQUALIS.

CHRYSOCOMA	**GOLDEY-LOCKS**			
1. Coma-aurea.	great ſhrubby.	C. G. H.	June, Sept.	G. tr.
2. ciliata.	ciliated.	———	July, Oct.	G. tr.
SANTOLINA	**LAVENDER COTTON**			
1. roſmarinifolia.	roſemary leaved.	Spain.	July, Sept.	G. tr.
ATHANASIA				
1. dentata.	tooth leaved.	C. G. H.	July, Auguſt.	G. tr.
2. crithmifolia.	ſamphire leaved.	———		G. tr.

P. SUPERFLUA.

TANACETUM	**TANSEY**			
1. ſuffruticoſum.	ſhrubby.	C. G. H.	May, Sept.	G. tr.
2. flabellifolium.	fan leaved.	———	May, Aug.	G. tr.
ARTEMISIA	**WORMWOOD**			
1. arboreſcens.	tree.	Levant.	June, Aug.	G. tr.
2. argentea.	ſilvery leaved.	Madeira.	June, July.	G. tr.
GNAPHALIUM	**EVERLASTING**			
1. arboreum.	tree.	C. G. H.	June, Aug.	G. tr.
2. grandiflorum.	great flowered,	———		G. ps.
3. ericoides.	heath leaved.	———	April, Aug.	G. tr.
4. igneſcens.	fiery.		June, Sept.	G. ps.

SYNGENESIA P. SUPERFLUA.

GNAPHALIUM	EVERLASTING			
5. patulum.	spreading.	C. G. H.		G. tr.
6. crassifolium.	thick leaved.		July.	G. tr.
7. maritimum.	sea			G. tr.
8. Orientale.	Eastern.	Africa.	April, Sept.	G. tr.
9. cymosum.	branching.			G. tr.
10. glomeratum.	cluster flowered.	C. G. H.	August.	G. an.
11. odoratissimum.	sweet scented.		April, Aug.	G. pe.
XERANTHEMUM				
1. retortum.	reflex leaved.	C. G. H.	July, Aug.	G. tr.
2. fulgidum.	yellow flowered.		Mar. Sept.	G. pe.
CONYZA				
1. candida.	hoary.	Crete.	June, July.	G. tr.
SENECIO	GROUNDSEL			
1. Pseudo-China.	Chinese.	China, E. India.		S. pe.
2. venustus.	wing leaved.	C. G. H.	July, Sept.	G. tr.
ASTER	STAR WORT			
1. fruticosus.	shrubby.	C. G. H.	April, June.	G. tr.
v. fl. alb.	white flowered ditto.			
2. dentatus.	tooth leaved.	New Holland.	May, June.	G. tr.
3. Cymbalaria.	ivy leaved.		June, July.	G. bi.
CINERARIA				
1. Amelloides.	Cape after.	C. G. H.	March, Sept.	G. tr.

SYNGENESIA P. SUPERFLUA.

CINERARIA	CINERARIA			
2. lanata.	woolly.	Canary Islands.	May.	G. tr.
3. geifolia.	kidney leaved.	C. G. H.	June, Aug.	G. tr.
4. populifolia.	poplar leaved.	Canary Islands.		G. tr.
v. fl. alb.	white flowered ditto.			G. tr.
5. cruenta.	purple leaved.	C. G. H.	Mar. May.	G. pr.
6. lobata.	lobed.		June, Aug.	G. tr.
7. malvaefolia.	mallow leaved.	Azores.	August.	G. tr.
INULA	INULA			
1. viscosa.	clammy.	S. of Europe.	Aug. Sept.	G. tr.
LEYSERA	LEYSERA			
1. gnaphaloides.	woolly leaved.	C. G. H.	July, Sept	G. tr.
RELHANIA	RELHANIA			
1. squarrosa.	cross leaved.	C. G. H.	May, June.	G. tr.
CHRYSANTHEMUM	CHRYSANTHEMUM			
1. lacerum.	cut leaved.	Canary Islands.	M. P. S.	G. tr.
2. frutescens.	shrubby.		June, Oct.	G. tr.
3. flosculorum.	tooth leaved.	Crete.	May, August.	G. tr.
4. pinnatifolium.	pinnated.	Madeira.		G. tr.
COTULA	COTULA			
1. stricta.	silvery.	C. G. H.	May, June.	G. tr.
SIGESBECKIA	SIGESBECKIA			
1. Orientalis.	East Indian.	E. India.	July, August.	S. an.

SYNGENESIA P. SUPERFLUA.

SIGESBECKIA	SIGESBECKIA				
2. flosculosum.	small flowered.	·	Peru.	June, July.	*S. an.*
VERBESINA	VERBESINA				
1. alata.	wing stalked.	·	S. America.	*M. P. S.*	*S. p.*
2. gigantea.	gigantic.	·	W. India.		*S. tr.*
BUPTHALMUM	OX-EYE				
1. fericeum.	silky.	-	Canary Islands.	May, July.	*G. tr.*
2. frutescens.	shrubby.	-	Virginia, Jamaica.	June, August.	*G. tr.*
3. maritimum.	sea.	-	Sicily.	July, Sept.	*G. p.*

P. FRUSTRANEA.

GORTERIA	GORTERIA				
1. ciliaris	ciliated.	-	C. G. H.	June, Aug.	*G. tr.*
2. fruticosa.	shrubby.	-	———	Aug. Sept.	*G. tr.*
3. rigens,	great flowered.	-	———	April, July.	*G. tr.*
CENTAUREA	CENTAURY				
1. ragusina.	white leaved.	-	Crete.	June, July.	*G. tr.*
2. sempervirens.	evergreen.	-	Spain, Portugal.	July, August.	*G. tr.*
3. Cineraria.	groundsel leaved.	-	Italy.	June, July.	*G. p.*
4. argentea.	silvery.	-	Crete.	———	*G. tr.*
5. spinosa.	prickly branched.	·	Candia.		*G. p.*

SYNGENESIA P. NECESSARIA.

P. NECESSARIA.

CALENDULA	MARYGOLD				
1. Tragus.	long leaved.	-	C. G. H.	May, June.	G. tr.
2. graminifolia.	grafs leaved.	-	——		G. pe.
3. fruticofa.	fhrubby.	-	——	June, July.	G. tr.
ARCTOTIS	ARCTOTIS				
1. fuperba.	fuperb.	-	C. G. H.	July, Auguft.	G. pe.
2. afpera.	rough.	-	——	July, Sept.	G. tr.
3. acaulis.	ftemlefs.	-	——	April, July.	G. pe.
4. plantaginea.	plantain leaved.	-	——	June, Auguft.	G. tr.
5. fcariofa.	fouthernwood leaved.	-	——	April, Auguft.	G. tr.
6. paleacea.	chaffy.	-	——	——	G. tr.
OTHONNA	RAGWORT				
1. pectinata.	wormwood leaved.	-	C. G. H	May, June.	G. tr.
2. abrotanifolia.	fouthernwood leaved.	-	——	March. May.	G. tr.
3. coronopifolia.	buck's horn leaved.	-	——	July, Sept.	G. tr.
4. arborefcens.	tree African.	-	——		G. tr.
OSTEOSPERMUM	OSTEOSPERMUM				
1. moniliferum.	poplar leaved.	-	C. G. H.	July, Auguft.	G. tr.
2. grandiflorum.	great flowered.	-	——	——	G. tr.

SYNGENESIA P. NECESSARIA.

HIPPIA				
1. frutescens.	shrubby.	C. G. H.	July, August.	G. tr.
ERIOCEPHALUS				
1. racemosus.	silver leaved.	C. G. H.	Mar. Aug.	G. tr.
2. Africanus.	cluster leaved.		———	G. tr.

POL. SEGREGATA.

ŒDERA				
1. prolifera.	stiff leaved.	C. G. H.	May, June.	G. tr.

MONOGAMIA.

LOBELIA				
1. pinifolia.	pine leaved.	C. G. H.	June, July.	G. tr.
2. triquetra.	tooth leaved.	Jamaica.	May, Sept.	G. pe.
3. longiflora.	long flowered.	C. G. H.	June, August.	S. pe.
4. minuta.	minute.	———	June, Sept.	G. pe.
5. Erinus.	small spreading.			G. pe.
6. pubescens.	downy.	———	May, Aug.	G. pe.

H h

CLASSIS XX.

GYNANDRIA DIANDRIA.

LIMODORUM				
1. Tankervilliæ.	Tankerville's.	- China.	Mar. Apr.	S. pe.
2. altum.	tall.	- W. India.	May, July.	S. pe.
EPIDENDRUM				
1. cochleatum.	shell flowered.	- W. India.	June, July.	S. pe.
2. ensifolium.	sword leaved.	- China, Japan.		S. pe.
3. aloides.	aloe-leaved.	- E India.		S. pe.
GUNNERA				
1. Perpensa.	marsh marygold leaved.	- C. G. H.	June.	G. pe.

TRIANDRIA.

FERRARIA				
1. undulata.	waved.	- C. G. H.	Apr. May.	G. pe.
2. pavonia.	spotted.	- Mexico.	May, June.	G. pe.

GYNANDRIA PENTANDRIA.

PENTANDRIA.

PASSIFLORA	PASSION FLOWER			
1. ferratifolia.	faw leaved.	W. India.	May, Oct.	S. tr.
2. maliformis.	apple fruited.	Dominica.		S. tr.
3. quadrangularis.	fquare.	Jamaica.	Aug. Sept.	S. tr.
4. laurifolia.	laurel leaved.	W. India.	June, July.	S. tr.
5. lunata.	crefcent leaved.			S. tr.
6. incarnata.	flefh coloured.	America.		S. pe.
7. fuberofa.	cork barked.	W. India.	Sept.	S. tr.
8. lutea.	yellow.	Jamaica, Virginia.	May, July.	S. pe.
9. glauca.	glaucous.	Cayenne.	Aug. Sept.	S. tr.
10. holofericea.	filky leaved.	Vera Cruz.	June, Sept.	S. tr.
11. punctata.	dotted.	Peru.	May, June.	S. tr.
12. heterophylla.	various leaved.	W. India.	June, Sept.	S. tr.
13. minima.	dwarf.	Curaffao.	July.	S. tr.
14. normalis.	Carthagena.	S. America.		S. tr.
15. murucuja.		Dominica.		S. tr.

H h 2

GYNANDRIA HEXANDRIA.

HEXANDRIA.

ARISTOLOCHIA	BIRTH-WORT				
1. sempervirens.	evergreen.	-	Crete.	May, June.	G. tr.
2. odoratissima.	sweet scented.	-	Jamaica.		S. tr.
3. caudata.	tailed.	-	America.		S. tr.
4. Pistolochia.			Spain, Narbonne.		G. pe.

DECANDRIA.

HELICTERES	SCREW TREE				
1. Isora.	twisted capsuled.	-	Jamaica.	June, July.	S. tr.

POLYANDRIA.

ARUM					
1. esculentum.	eatable.	-	Indies, China.		S. pe.
2. Colocasia.	Egyptian.	-	Levant.		S. pe.
3. bicolorum.	two coloured.			June, July.	S. pe.
4. trilobum.	three leaved.	-	Ceylon.	May, June.	S. pe.

GYNANDRIA POLYANDRIA.

ARUM	**ARUM**			
5. seguinum.	dumb cane.	W. India.	March,	S. tr.
6. venosum.	purple flowered.	-		S. ps.
CALLA	**CALLA**			
1. Æthiopica.	Ethiopian.	Africa.	Feb. May.	G. pe.
POTHOS	**POTHOS**			
1. cordata.	heart leaved.	W. India.	April, May.	S. pe.
2. lanceolata.	spear leaved.	——	——	S. pe.

CLASSIS XXI.

MONOECIA MONANDRIA.

CASUARINA	**CASUARINA**			
1. equisetifolia.	horse tail.	New Holland.	October.	G. tr.
2. stricta.	upright.	——	Jan. March.	G. tr.
3. tortulosa.	cork-bark.	——		G. tr.

MONOECIA TRIANDRIA.

TRIANDRIA.

ZEA	INDIAN CORN			
1. Mays.	common.	America.	June, July.	*G. an.*
COIX	JOB'S TEARS			
1. Lacryma-Jobi.	common.	E. India.	June, July.	*S. pe.*
HERNANDIA	JACK IN A BOX			
1. fonora.	target leaved.	W. India.		*S. tr.*

TETRANDRIA.

URTICA	NETTLE			
1. nivea.	fnowy.	China.	Aug. Sept.	*G. tr.*

PENTANDRIA.

XANTHIUM	XANTHIUM			
1. fruticofum.	fhrubby.	Peru.	July.	*G. tr.*
PARTHENIUM	PARTHENIUM			
1. Hyfterophoros.	cut leaved.	Jamaica.	July, Aug.	*G. an.*

MONOECIA POLYANDRIA.

POLYANDRIA.

BEGONIA				
climbing sorrel.	-	Indies.		*S. p.*
1. obliqua.				
POTERIUM BURNET				
spiny shrubby.	-	Levant.	April, Aug.	*G. tr.*
1. spinosum.				
smooth shrubby.	-	Canary Islands.	March, April.	*G. tr.*
2. caudatum.				

MONADELPHIA.

CYPRESS				
juniper leaved.	-	C. G. H.	April, May.	*G. tr.*
1. juniperoides.				
CROTON				
tallow tree.	-	China.	September.	*G. tr.*
3. sebiferum.				
JATROPHA PHYSIC NUT				
multifid.	-	S. America.	June, August.	*S. tr.*
1. multifida.				
stinging,			May, July.	*S. tr.*
2. urens.				
RICINUS				
PALMA CHRISTI, or *Castor oil tree.*				
common.	-	Indies.	August.	*S. tr.*
1. communis.				
HIPPOMANE MANCHINEEL TREE				
laurel leaved.	-	W. India.		*S. tr.*
1. biglandulosa.				
HURA SAND BOX TREE				
common.	-	W. India.		*S. tr.*
1. crepitans.				

CLASSIS XXII.

DIOECIA TRIANDRIA.

RESTIO	C. G. H.	July.	G. tr.
1. Elegia. rush leaved.			

TETRANDRIA.

MYRICA, or *Candle berry myrtle.*			
1. Faya. Azorian.	Azores.	June, July.	G. tr.
2. incisa. cut leaved.	C. G. H.	Auguſt.	G. tr.
3. cordifolia. heart leaved.	————	———	G. tr.
4. quercifolia. oak leaved.	————	———	G.

PENTANDRIA.

PISTACHIA PISTACHIA TREE	S. of Europe.	May.	G. tr.
1. Lentiſcus. Maſtick.			

HEXANDRIA.

SMILAX SMILAX Ceylon.	E. India.		S. tr.
1. Zeylanica.			

DIOECIA DECANDRIA.

DECANDRIA.

CARICA 1. Papaya.	PAPAW TREE common.	- Indies.	July.	*S. tr.*
KIGGELARIA 1. Africana.	KIGGELARIA African.	C. G. H.	May, June.	*G. tr.*
SCHINUS 1. molle.	SCHINUS wing leaved.	- Peru.	July, August.	*G. tr.*

POLYANDRIA.

CLIFFORTIA 1. ilicifolia.	CLIFFORTIA ilex leaved.	- C. G. H.	June, July.	*G. tr.*
2. obcordata.	heart leaved.	-		*G. tr.*
FLACOURTIA 1. Bamontchi.	FLACOURTIA shining leaved.	- Madagascar.	June, July.	*S. tr.*

MONADELPHIA.

TAXUS 1. elongata.	YEW TREE long leaved.	- C. G. H.	July.	*G. tr.*

I i

DIOECIA MONADELPHIA.

CISSAMPELOS 1. Capensis.	oval leaved.	- C. G. H.		*G. tr.*

SYNGENESIA.

RUSCUS 1. androgynus.	**BUTCHER'S BROOM** climbing.	- Canary Islands.	May, July.	*G. tr.*
BRYONIA 1. grandis.	great flowered.	- Indies.		*S. pe.*

GYNANDRIA.

CLUYTIA 1. pulchella.	broad leaved.	- C. G. H.	March, June.	*G. tr.*
2. alaternoides.	narrow leaved.	- ———	Dec. Mar.	*G. tr.*
3. daphnoides.	daphnoides.	- ———		*G. tr.*
4. polygonoides.	hoary leaved.	———		*G. tr.*

CLASSIS XXIII.

POLYGAMIA MONOECIA.

MUSA				
1. fapientum.	Banana.	- W. India.	July, Nov.	*S. p.*
2. coccinea.	fcarlet.	- E. India.	Jan. April.	*S. p.*
GOUANIA	CHAW STICK	-		*S. tr.*
1. domingenfis.	common.			
HOLCUS	HOLCUS			
1. bicolor.	two coloured.	Perfia.	July.	*S. an.*
2. Sorghum.	Indian millet.	- E. India.		*S. an.*
PARIETARIA	PELLITORY			
1. arborea.	tree.	- Canary Iflands.	March, May.	*G. tr.*
TERMINALIA	TERMINALIA			
1. angustifolia.	narrow leaved.	- E. India.		*S. tr.*
OPHIOXYLUM	OPHIOXYLUM			
1. ferpentinum.	red flowered.	- Indies.	May, June.	*S. tr.*
MIMOSA	MIMOSA			
1. verticillata.	whorl leaved.	- New Holland.	March, May.	*G. tr.*
2. ulicina.	furze leaved.	- ————	————	*G. tr.*
3. enfifolia.	fword leaved.	- ————	————	*G. tr.*
4. falcata.	fickle leaved.	- ————	————	*G. tr.*
5. myrtifolia.	myrtle leaved.	- ————	————	*G. tr.*

POLYGAMIA MONOECIA.

MIMOSA

6. virgata.	long twigged.	New Holland.	July, August.	S. tr.
7. suaveolens.	sweet smelling.	-	March, May.	G. tr.
8. julibrissin.	tree.	-	August.	G. tr.
9. farnassana.	sponge tree.	Levant.	June, Aug.	S. tr.
10. pudica.	humble plant.	W. India.	M. P. S.	S. an.
11. purpurea.	soldier wood.	-		S. tr.
12. sensitiva.	sensitive plant.	-	M. P. S.	S. tr.
13. latisiliqua.	broad podded.	W. India.		S. tr.
14. cornigera.	cuckold tree.	S. America.		S. tr.
15. Intsia.	angular stalked.	E. India.		S. tr.
16. punctata.	large smooth.	America.		S. tr.
17. peregrina.	-	-		S. tr.
18. tamarindifolia.	tamarind leaved.	S. America.		S. tr.
19. Inga.	-	-		S. tr.
20. Ceratonia.	-	Ceylon.		S. tr.
21. nodosa.	-	-		S. tr.

DIOECIA.

DIOSPYRUS — DATE PLUMB

1. Kaki.		Japan.		S. tr.

POLYGAMIA TRIOECIA.

			Mar. Apr.	S. tr.
PISONIA				
1. aculeata.	prickly.	China.		S. tr.

TRIOECIA.

CERATONIA	CAROB TREE, or *St. John's bread*,			
1. Siliqua.	common.	- Sicily, Levant.		G. tr.
FICUS	FIG TREE			
1. religiofa.	poplar leaved.	· E. India.	April.	S. tr.
2. Benghalenfis.	Bengal.	· S. America.		S. tr.
3. pedunculata.	willow leaved.	· W. India.		S. tr.
4. virens.	round fruited.	· E. India.		S. tr.
5. Indica.	Indian.	·		S. tr.
6. venofa.	vein leaved.	·		S. tr.
7. coftata.	rib leaved.	·		S. tr.
8. benjamina,	oval leaved.	·		S. tr.
9. coriacea.	leathery.	·		S. tr.
10. ftipulata.	trailing,	· China, Japan.		S. tr.

CLASSIS XXIV.

CRYPTOGAMIA FILICES.

PTERIS	**BRAKE**			
1. longifolia.	long leaved.	-	W. India.	July, Sept.	*S. pe.*
2. arguta.	sharp notched.	-	Arabia, Madeira, C. G. H.	Aug. Sept.	*G. pe.*
3. ferrulata.	various leaved.	-	E. India, China.	August.	*S. pe.*
ASPLENIUM	**SPLEEN-WORT**			
1. Hemionitis.	mule's fern.	-	Madeira.	June, July.	*G. pe.*
POLYPODIUM	**POLYPODY.**			
1. aureum.	golden.	-	Jamaica.	March, July.	*S. pe.*
2. trifoliatum.	three leaved.	-	W. India.		*S. pe.*
3. asplenifolium.	spleen wort leaved.	-	S. America.		*S. pe.*
ADIANTUM	**MAIDEN HAIR**			
1. reniforme.	kidney leaved.	-	Madeira.	June, Sept.	*G. pe.*
BLECNUM	**BLECNUM**			
1. radicans.	rooting leaved.	-	Madeira.	*M. P. S.*	*G. pe.*
2. Australe.	Cape.	-	C. G. H.		*G. pe.*
THRICHOMANES	**HARE'S FOOT FERN**			
1. Canariense.	Canary.	-	Portugal, Canaries.	*M. P. S.*	*G. pe.*
DICKSONIA	**DICKSONIA**			
1. culcita.	shining leaved.	-	Madeira, Azores.	June, July.	*G. pe.*

PALMAE.

RHAPIS				
1. flabelliformis.	fan, or *ground ratan*.	China, Japan.	Auguſt.	S. tr.
CHAMAEROPS				
1. humilis.	dwarf.	S. of Europe.	May, July.	G. tr.
CYCAS				
1. revoluta.	narrow leaved.	China, Japan.	Aug.	S. tr.
PHOENIX				
1. dactylifera.	DATE PALM common.	Levant.		S. tr.
ZAMIA				
1. debilis.	long leaved.	W. India.	July, Aug.	S. tr.
2. integrifolia.	dwarf.	E. Florida.		S. tr.
BORASSUS				
1. flabelliformis.	PALMAIRA TREE common.	Coaſt of Coromandel.		S. tr.
CORYPHA				
1. umbraculifera.	FAN-PALM great.	E. India.		S. tr.

FINIS.

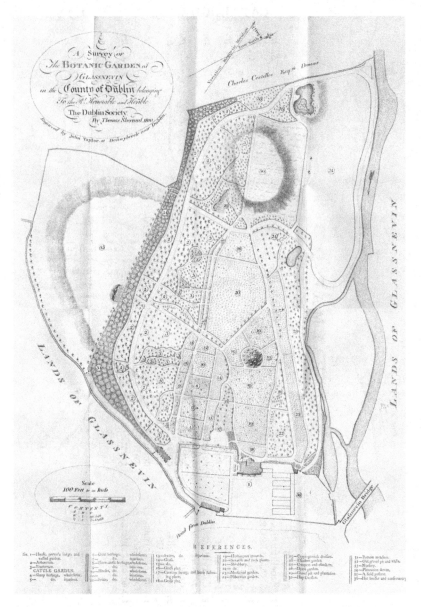

The material originally positioned here is too large for reproduction in this reissue. A PDF can be downloaded from the web address given on page iv of this book, by clicking on 'Resources Available'.

CATALOGUE

OF

PLANTS

IN THE

ARBORETUM,	GRAMINA VERA,
FRUTICETUM,	HORTUS TINCTORIUS,
HERBARIUM,	HOT & GREEN HOUSES

OF THE

DUBLIN SOCIETY'S BOTANIC GARDEN,

AT GLASNEVIN.

NOMINA SI NESCIS, PERIT ET COGNITIO RERUM.

PRINTED BY GRAISBERRY AND CAMPBELL,
NO. 10, BACK-LANE.

1802.

CATALOGUE

OF No. 2.

*In the Map of the Dublin Society's Botanic Garden at Glaf-
nevin, containing the ARBORETUM, or Systematic
arrangement of Trees.*

☞ NOTE. The Claffes are diftinguifhed by tall Marks,
each clafs in a feparate PLOT, furrounded by a Gravel
Walk, and the Marks of the Species and Varieties, are in
the Form of CROSSES, with BLACK and RED Letters on
a WHITE GROUND, and the Croffes begin about the S. E.
End of the Garden.

PENTANDRIA.

ULMUS campeftris *v. vul.*	Common Elm
——— *v. fol. aur. var.*	Gold blotched ditto
——— *v. fol. arg. var.*	Silver blotched ditto.
——— *v. ftriƌa.*	Cornifh Elm.
——— *v. latifol.*	Welch ditto
——— *v. glab.*	Smooth ditto
——— *v. fol. aur. var.*	Gold blotched ditto
——— *v. fungos.*	Dutch ditto
—— Americana *v. rub.*	Red American ditto
——— *v. alb.*	White ditto
——— *v. pendul.*	Weeping ditto
——— pumila.	Dwarf ditto
——— nemoralis.	Horn-beam leaved ditto
—— non *defcript. fol. rugos.*	Rough leaved ditto

HEPTANDRIA.

Æfculus Hippo-caftanum	Common horfe chefnut
—— *v. fol. aur. var.*	Gold blotched ditto

B   Æfculus

Æfculus Hippo-caftanum *v. fol.* Silver blotched .horſe chefnut
    arg. var.

——————— flava.   Yellow flowered ditto
——————— Pavia.   Scarlet ditto
——————— *nova ſpecies.*

DECANDRIA.

Sophora tetraptera.   Winged podded Sophora.

ICOSANDRIA.

Prunus Padus.   Common bird cherry.
———————————— *v. minor.*   Dwarf ditto
———— rubra.   Cornifh ditto
———— Virginiana.   American ditto
———— Caroliniana.   Evergreen ditto
———— Luſitanica.   Portugal laurel
———————————— v. *fol. aur. var.* Gold blotched ditto
———— Lauro-ceraſus.   Common laurel
———————————— *v. anguſt.*   Slender leaved ditto
———————————— *v. fol. aur. var.* Gold blotched ditto
———— Mahaleb.   Perfumed cherry
———— Armeniaca.   Common apricot tree
———————————— *v. fol. aur. mac.* Gold blotched ditto
———— pumila.   Dwarf Canadian cherry
———— Ceraſus.   Cultivated ditto
———————————— *v. fol. aur. var* Gold blotched ditto
———————————— *v. virg. fol. pen.* Weeping leaved ditto
———————————— *v. ramos. pen.*   Weeping branched ditto
———————————— *v. fl. plen.*   Double flowered ditto
———— avium.   Small fruited ditto
———— Penſylvanica.   Upright ditto
———— nigra.   Black ditto
———— domeſtica.   Common plum tree
———————————— *v. fol. aur. mac.* Gold blotched ditto
———————————— *v. fol. arg.*   Silver blotched ditto
———— inſititia.   Common Bullace tree
———— ſpinoſa.   Sloe tree

                                                  Prunus

Prunus Sibirica.   Siberian Cherry
——— falicina.

POLYANDRIA.

Tilia Europea.   Lime tree
——————— *v. corallina.*   Red twigged ditto
——————— *v. parvifol.*   Small leaved ditto
——— Americana,   Broad leaved ditto
——— alba.   White ditto
Liriodendron Tulipifera.   Tulip tree
Magnolia grandiflora.   Common laurel leaved Magno-
——————— *v. obovata.*   Broad leaved ditto   [lia
——————— *v. lanceol.*   Long leaved ditto
——— glauca.   Deciduous fwamp ditto
——————— *v. longif.*   Evergreen ditto
——— acuminata.   Blue ditto
——— tripetala.   Umbrella ditto

DIADELPHIA.

Robinia Pfued-Acacia.   Common Acacia.

MONOECIA.

Betula alba.   Common Birch
——— *v. pend.*   Weeping ditto
——— populifolia.   Poplar leaved birch tree
——— nigra.   Black ditto
——— papyracea.   Paper ditto
——— lenta.   Soft ditto
——— excelfa.   Tall ditto
——— nana.   Smooth dwarf ditto
——— pumila.   Hairy ditto
——— oblongata.   Oblong leaved ditto
——————— *v. ellipt.*   Oval leaved ditto
——— Alnus.   Common alder tree
——————— *v. pinnatif.*   Cut leaved ditto

Betula ſerulata.	Notched leaved ditto.
—— incana.	Glaucous leaved ditto
———— *v. angul.*	Elm leaved ditto
—— Americana.	American ditto
—— Sibrica.	Siberian ditto
—— criſpa.	Curled leaved ditto
Morus alba.	White mulberry tree
—— nigra.	Common ditto
—— papyrifera.	Paper ditto
—— rubra.	Red ditto
Quercus Phellos	Common willow leaved oak tree
———— *v. ſericea*	Dwarf ditto.
—— Ilex.	Common evergreen ditto
———— *v. ſerrata.*	Saw leaved ditto
———— *v. oblong.*	Long leaved ditto
—— gramuntia.	Holly leaved ditto
—— Suber.	Cork tree
—— coccifera.	Kermes oak tree
—— virens.	Live ditto
—— Prinos.	Broad cheſnut leaved ditto.
———— *v. oblong.*	Long ditto
—— aquatica.	Common water oak tree
———— *v. heteroph.*	Various leaved ditto
———— *v. elong.*	Long leaved ditto
———— *v. indiv.*	Entire leaved ditto
———— *v. attenuata.*	Narrow leaved ditto
—— nigra.	Black oak tree
—— rubra.	Broad leaved red campion do.
———— *v. coccinea.*	Scarlet ditto
———— *v. mont.*	Mountain ditto
—— diſcolor.	Downy leaved ditto
—— alba.	White ditto
—— Eſculus.	Small prickly cup'd ditto
—— Robur.	Female oak
———— *v. ſeſſil.*	Common ditto

Quercus

Quercus Robur *v. humilis.*	Dwarf ditto
———— *v. fol. perrennant.*	Turner's ditto
———— *v. fol. arg. var.*	Silver blotched ditto
———— Ægilops.	Great prickly cup'd ditto
———— Cerris.	Common Turkey ditto
———— *v. bullat.*	Rough leaved ditto
Quercus Cerris.	Narrow leaved oak
———— *v. fol. perennaut.*	Bagnal ditto
———— *v. fol. subtus alb.*	Luccombe ditto
———— *v. fol. pinnatif.*	Winged leaved ditto
———— tinctoria.	Dyer's oak
Salisburia adiantifolia *vulg. Gink-*go.	Maiden hair tree
Juglans regia.	Common wallnut tree
———— alba.	White hickery
———— nigra.	Black wallnut tree
———— cinerea.	Shell bark ditto
———— angustifolia.	Ilionoise ditto
Fagus Castanea	Narrow leaved common chesnut
———— *v. fol. aur var.*	Gold blotched ditto
———— pumila.	Dwarf chesnut tree
———— ferruginea.	American beech tree
———— sylvatica.	Common ditto
———— *v. purp.*	Purple ditto
———— *v. cuprifol.*	Copper leaved ditto
Carpinus Betulus.	Common hornbeam tree
———— *v. incisa.*	Cut leaved ditto
———— *v. fol. aur. var.*	Gold blotched ditto
———— *v. quercif.*	Oak leaved ditto
———— Ostrya.	Hop ditto
———— Virginiana.	Flowering ditto
Plantanus Orientalis.	Oriental plane tree
———— *v. acerifol.*	Spanish ditto
———— *v. undul.*	Wave leaved ditto
———— Occidentalis.	American ditto
———— *v. fub. glab.*	Smooth ditto

Pinus

Pinus fylveftris.	Common Scotch firr
———— v. fol. var.	Variegated ditto
———— v. Tatarica.	Tartarian pine tree
———— v. mont.	Mountain, or Mugho ditto
———— v. divaricat.	Hudfon's Bay ditto
———— v. marit.	Sea ditto
——— Pinafter.	Pinafter, or clufter ditto
——— inops.	Jerfey ditto
——— refinofa.	American pitch tree
——— Halepenfis.	Aleppo pine tree
——— pinea.	Stone ditto
——— tœda	Frankincenfe ditto
———— v. rigida.	Three leaved Virginian ditto
———— v. variab.	Two and three leaved ditto
———— v. alopecur.	Foxtail ditto
——— paluftris.	Swamp ditto
——— Cembra.	Siberian ftone ditto
——— Strobus.	Weymouth pine
——— Cedrus.	Cedar of Lebanon
——— pendula.	Black larch tree
——— Larix.	Common white ditto
——— picea.	Silver firr tree
——— balfamea.	Balm of Gilead ditto
——— Canadenfis.	Hemlock fpruce firr tree
——— nigra.	Black ditto
——— Abies.	Norway ditto
——— alba.	White fpruce firr tree

DIOECIA.

Salix alba.	White willow.
——— purpurea.	Purple ditto
——— pentandra.	Sweet ditto
——— hermaphroditica.	Shining ditto
——— Babylonica.	Weeping ditto
——— triandra.	Smooth ditto

Salix

Salix fragilis.	Crack ditto
—— amygdalina.	Almond leaved ditto
—— haftata.	Halbert leaved ditto
Populus alba.	Common white poplar
—— *v. nivea.*	Great white Abele tree
—— tremula.	Afpen tree
—— —— *v. pend.*	Weeping ditto
—— nigra.	Black poplar
—— Chefterienfis.	Chefter ditto
—— dilatata.	Lombardy, or Po poplar
—— balfamifera.	Common Tacamahac ditto
—— *v. nova.*	
—— candicans.	Heart leav'd ditto
—— lævigata.	Smooth ditto
—— monolifera.	Canadian ditto
—— Græca.	Athenian ditto
—— Curtifia.	
—— heterophylla.	Various leav'd ditto
—— angulata.	Carolina ditto
—— Canadenfis.	Canadian ditto
Taxus baccata.	Common yew tree
—— elongata.	Long leav'd ditto
—— nucifera.	Nut bearing ditto

POLYGAMIA.

Acer Tataricum.	Tartarian maple
——Pfeudo—Planatus.	Great maple, or fycamore
—— *v. fruct. coccin.*	Scarlet berried ditto
—— *v. fol. aur. var.*	Gold blotched ditto
—— *v. fol. arg. var.*	Silver blotched ditto
—— rubrum	Scarlet flowered maple
—— *v. pallidum.*	Pale flowered ditto
—— faccharinum.	Sugar ditto
—— Platanoides.	Norway ditto
—— *v. laciniato.*	Cut leav'd ditto

Acer

Acer montanum.	Mountain ditto
—— Penſylvanicum.	Penſylvanian ditto
—— campeſtre.	Common ditto
—————— *var. fol. arg. var.*	Silver blotched ditto
—— Opalus.	Italian ditto
—— hybridum.	Hybrid ditto
—— Monſpeſulanum.	Montpellier ditto
—— Creticum.	Cretan ditto
—— Negundo.	Aſh leav'd ditto
Gleditſia triacanthos.	Three thorned acacia
————— *v. monos.*	Single ſeeded, or water ditto
————— *v. horrida.*	Strong ſpined ditto
Fraxinus excelſior.	Common aſh tree
————— *v. pend.*	Weeping ditto
————— *v. diverſif.*	Various leav'd ditto
————— *v. integrif.*	Entire leav'd ditto
————— *v. fol. criſp.*	Curled leaved ditto
————— *v. pumila.*	Dwarf ditto
————— *v. fol. aur.*	Gold blotched ditto
————— *v fol. arg.*	Silver blotched ditto
————— *v. cortice ſtriat.*	Striped barked ditto
————— *v. nigricant.*	Chineſe ditto
——— Ornus.	Flowering ditto
——— rotundifolia.	Manna ditto
——— Americana.	American ditto
——— *v. alb.*	American white ditto

No. 3.

Containing the FRUTICETUM *or* SYSTEMATIC ARRANGE-
MENT *of* SHRUBS.

☞ NOTE.—The Claſſes are diſtinguiſhed by tall Marks, each
Claſs in a ſeparate PLOT, ſurrounded by a Gravel Walk,
and the Marks of the Species and Varieties are in the Form
of CROSSES, with BLACK and RED LETTERS on a WHITE
GROUND, and the CROSSES begin about the Eaſt End of
the Garden.

DIANDRIA,

Jaſminum fruticans.	Common yellow jaſmine
———— *v. fol. aur. var.*	Gold blotched ditto
———— *v. Luſitan.*	Portugal ditto
——— humile.	Italian yellow ditto
——— officinale.	Common white ditto
———— *v. fol. aur. var.*	Gold blotched ditto
———— *v fol. arg. var.*	Silver blotched ditto
Liguſtrum vulgare.	Common privet
———— *v. fol. aur. var.*	Gold ſtriped ditto
———— *v. fol. arg. var.*	Silver ſtriped ditto
———— *v. ſempervir.*	Evergreen ditto
Phillyrea media	Phillyrea
———— *v. virgata.*	Long branch'd ditto
———— *v. pend.*	Drooping ditto.
———— *v. oleæfolia*	Olive leav'd ditto
———— *v. buxifol.*	Box leav'd ditto

C

Phillyrea

Phillyrea angustifolia.	Common narrow leav'd ditto
———————— v. rosmarinif.	Rosemary leav'd ditto
———————— v. brachiata.	Dwarf ditto
——— latifolia.	Smooth broad leav'd ditto
———————— v. fol. varieg.	Variegated ditto
———————— v. spinosa.	Prickly broad leav'd ditto
———————— v. obliqua.	Ilex leav'd ditto.
———————— v. fol. aur. var.	Gold blotched ditto
Olea Europea.	Common olive
——— Americana.	American ditto
Chionanthus Virginica.	Broad leav'd fringe tree
——————— v. angust.	Narrow leav'd ditto
Syringa vulgaris v. cærul.	Common blue lilac
——————— v. violac.	Common purple ditto
——————— v. alb.	Common white ditto
——————— v.	William's ditto
——— Persica.	Blue Persian ditto
——————— v. alb.	White Persian ditto
——————— v. laciniat.	Cut leav'd Persian ditto
Veronica decussata.	Cross leav'd speedwell
Rosmarinus officinalis.	Common rosemary
——————— v. fol. aur. var.	Gold striped ditto
Salvia Cretica.	Cretan sage
——— officinalis.	Garden ditto
——————— v. fol. aur. var.	Gold blotched ditto
——————— v. fol. aur. var.	Silver blotched ditto
——— triloba.	Three lobed ditto

TRIANDRIA.

Cneorum tricoccum.	Widdow wail

TETRANDRIA.

Cephalanthus Occidentalis.	American button wood
Mitchella repens.	Creeping Mitchella
Buddlea globosa.	Round headed Buddlea
Callicarpa Americana.	Johnsonia

Plantage

Plantago Cynops.	Shrubby plaintain
Cornus florida.	Great flower'd dogwood
———— *v. fol. aur. var.*	Gold blotched ditto
——— mafcula.	Cornelian ditto
——— alba.	White berried ditto
——— fanguinea,	Common ditto
———— *v. fol. aur. var.*	Gold blotched ditto
——— fericea.	Blue berried ditto
——— paniculata.	New Holland ditto
——— alternifolia.	Red twig'd alternate leav'd do.
———— *v. virefcens.*	Green twig'd ditto
——— circinata.	Siberian ditto
Ptelea trifoliata.	Shrubby trefoil
Elæagnus anguftifolia.	Narrow leav'd oleafter
Hamamelis Virginica.	Witch hazel
Ilex Aquifolium.	Common holly
———— *v. heterophylla.*	Various leav'd ditto
———— *v. craffifol.*	Thick leav'd ditto
——— *v. recurva.*	Slender leav'd ditto
——— *v. anguft.*	Maple leav'd ditto
——— *v. inermis.*	True Dutch ditto
——— *v. fruct. flav.*	Yellow berried ditto
——— *v. fruct. alb.*	White berried ditto
——— *v. fol. arg.*	Silver blotched ditto
——— *v. fol. aur.*	Gold blotched ditto
——— *v. ferox.*	Hedge-hog ditto
——— *v. ferox fol. aur.*	Gold blotched ditto
——— *v. ferox fol. arg.*	Silver blotched ditto
——— opaca.	Carolina ditto
——— prinoides.	Deciduous ditto
——— Caffine.	Broad leav'd ditto
——— *v. anguft.*	Narrow leav'd ditto
——— *v. minor.*	Small ditto
——— *v. dentata*	Toothed ditto
——— vomitoria.	South fea tea
——— Perado.	Thick leav'd fmooth holly

PENTANDRIA

PENTANDRIA.

Azalea nudiflora.	Deep scarlet Azalea
—————— v. rutilans.	Deep red ditto
—————— v. carnea.	Pale red ditto
—————— v. alba.	Early white ditto
—————— v. bicolor.	Red and white ditto
—————— v. papilionacea.	Variegated ditto
—————— v. phœnicea.	
—————— v. purpurescens.	
—————— v. crispa.	
—————— v. fastigiata.	
—————— v. partita.	Downy Azalea
———— viscosa.	Common white ditto
—————— v. vittata.	Striped flowered ditto
—————— v. fissa.	Narrow petal'd white ditto
—————— v. floribund.	Clustered flowered ditto
—————— v. glauca.	Glaucous ditto
—————— v. salicifol.	Willow leav'd ditto
———— Pontica.	Yellow Azalea
——— procumbens.	Procumbent ditto
Lonicera dioica.	Glaucous honeysuckle
——— Caprifolium.	Italian early white ditto
—————— v. rub.	Italian early red ditto
——— sempervirens.	Great trumpet ditto
—————— v. minor.	Small trumpet ditto
——— grata.	Evergreen ditto
——— implexa.	Minorca ditto
——— Periclymenum.	Common woodbine
—————— v. serotinum.	Late red woodbine
—————— v. Belgicum.	Dutch ditto
—————— v. quercifol.	Oak leav'd ditto
—————— v. quercif. fol. var.	Silver striped ditto
——— Nigra.	Black berried upright ditto
——— Tatarica.	Tartarian ditto
——— Xylosteum.	Fly ditto

Lonicera

Lonicera Pyrenaica.	Pyrenian upright ditto
———— alpigena.	Red berried ditto
——— cærulea.	Blue berried ditto
——— Symphoricarpos.	Shrubby St. Peter's wort
——— Dierville.	Yellow flowered upright honey-
——— Balearica.	Minorca ditto [fuckle
Solanum dulcamara.	Common woody nightſhade
——————— *v. fol. varieg.*	Silver ſtriped ditto
Lycium Barbarum.	Willow leav'd box-thorn
——————— *v. Chinenſe.*	Chineſe ditto
——— Europeum.	European ditto
Bumelia tenax.	} Silvery leav'd iron wood
Sideroxylon tenax.	
——————— Lycioides.	Willow leav'd ditto
Rhamnus catharticus.	Purging buck-thorn
——— infectorius.	Yellow berried ditto
——— ſaxatilis.	Rock ditto
——— Frangula.	Berry bearing alder
——— latifolius.	Azorian Rhamnus
——— alnifolius.	Alder leav'd ditto
——— Alaternus.	Common Alaternus
——————— *v. fol. arg.*	Silver blotched ditto
——————— *v. anguſt.*	Jagged leav'd ditto
——— Sibiricus.	Siberian ditto
——— volubilis.	Twining Rhamnus, or ſupple-
——— Paliurus.	Common Chriſt's thorn [jack
Ceanothus Americanus.	New Jerſey tea
Celaſtrus ſcandens.	Climbing ſtaff tree
Myrſine Africana.	African Myrſine
——— retuſa.	Tamaja
Euonymus Europea.	Common ſpindle tree
——— *v. fruct. alb.*	White fruited ditto
——— *v. fruct. pallid.*	Pale fruited ditto
——— *v. fol. aur. var.*	Gold blotched ditto
——— latifolia.	Broad leav'd ditto
——— verrucoſa.	Warted ditto

Euonymus

Euonymus atropurpurea.	Purple flowered ditto
—— Americana.	Evergreen ditto
Itea Virginica.	Virginean Itea
Ribes rubrum.	Red currants
——— v. fol. arg. var.	Silver blotched red ditto
—— album.	White currant
——— v. fol. arg. var.	Silver blotched ditto
—— spicatum.	
—— alpinum.	Alpine ditto
——— v. fol. aur. mac.	Gold blotched ditto
—— nigrum.	Black currants
—— glandulosum.	Glandulous ditto
—— floridum.	American ditto
—— diacantha.	Two spined gooseberry
—— Grossularia.	Common ditto
—— Uva-Crispa.	Smooth fruited ditto
—— oxyacanthoides.	Hawthorn leav'd ditto
——— v. fol. aur. var.	Gold blotched ditto
——— v. fol. arg. var.	Silver ditto
—— Cynosbati.	Prickly fruited currant.
—— petræum.	
Hedera Helix.	Common ivy
——— v. fol. arg. var.	Silver blotched ditto
——— v. fol aur. var.	Gold blotched ditto
—— latifolia.	Broad leav'd ditto
Vitis vinifera.	Common vine
—— Labrusca.	Downy leaved ditto
—— vulpina.	Fox grape
—— laciniosa.	Parsley leav'd vine
—— hederacea.	Ivy leav'd ditto
—— arborea.	Pepper ditto
Vinca minor v. fl. alb.	White flowered small periwinkle
——— v. fl. cærul.	Blue flowered ditto
——— v. fl. purp.	Purple flowered ditto
——— v. fl. subpleno.	Semidouble flowered small ditto
——— v. fol. aur. alb.	Gold blotched white flowered do.

Vinca

Vinca minor *v. fol. arg. alb.*	Silver blotched white flow. do.
— —— *v. fol. aur. cærul.*	Gold blotched blue flowered do.
———— *v. fol. arg. cærul.*	Silver blotched blue flowered do.
—— major.	Great blue flowered ditto
Periploca Græca.	Virginian filk
Salfofa fruticofa.	Stone crop tree
Bupleurum fruticofum.	Shrubby hare's ear
Rhus Coriaria.	Elm leaved fumach
—— Typhinus.	Virginian ditto
—— glaber.	Scarlet ditto
—— elegans.	Carolina ditto
—— vernix.	Varnifh ditto
—— radicans.	Common upright poifon oak
———— *v. lucidus.*	Small leav'd ditto
—— Toxicodendron.	Trailing ditto
—— aromaticus.	Aromatic fumach
—— fauveolens	Sweet ditto
—— cotinus.	Venetian ditto
—— copallinus.	Lentifcus leav'd ditto
Viburnum Tinus.	Laureftinus
———— *v. lucidum.*	Shining ditto
———— *v. virgatum.*	Common ditto
———— *v. ftrict.*	Upright ditto
———— *v. fol. aur. var.*	Gold blotched ditto
———— *v. fol. arg. var.*	Silver blotched ditto
—— nudum.	Oval leav'd Viburnum
—— caffinoides.	Thick leav'd ditto
—— nitidum.	Shining leav'd ditto
—— lævigatum.	Caffioberry bufh
—— prunifolium.	Plum leav'd Viburnum
—— Lentago.	Pear leav'd ditto
—— dentatum.	Shining tooth-leav'd ditto
———— *v. pub.*	Downy tooth-leaved ditto
———— *v. fol. aur. var.*	Gold blotchod ditto
—— acerifolium.	Maple leav'd ditto
—— Lantana.	Common wayfaring tree
	Viburnum

Viburnum Lantana *v. grand.*	Large leav'd ditto
———— *v. fol. aur. var.*	Gold blotched ditto
——— Opulus.	Marſh gilder roſe
———— *v. American.*	Red twig'd ditto
———— *v. roſeo.*	Snow ball ditto
——— Sibiricum.	Siberian ditto
Sambucus nigra.	Common elder
———— *v. viridis.*	Green berried ditto
———— *v. alb.*	White berried ditto
———— *v. fol. aur. var.*	Gold blotched ditto
———— *v. fol. arg. var.*	Silver blotched ditto
———— *v. laciniata.*	Parſley leav'd ditto
———— *v. fol. aur. var.*	Gold blotched ditto
——— Canadenſis.	Canadian ditto
——— racemoſa.	Red berried ditto
Staphylea pinnata.	Five leav'd bladder nut
——— trifolia.	Three leav'd ditto
Tamarix Gallica.	French tamariſk
——— Germanica.	German ditto
Xylophylla ramiflora.	Siberian Xylophylla
Aralia ſpinoſa.	Angelica tree
Zanthorhiza apiifolia.	Parſley leav'd Zanthoriza
Yucca glorioſa.	Adam's needle
——— Draconis.	Drooping leav'd ditto
——— filamentoſa.	Thready ditto
Prinos glabra.	Evergreen winterberry
——— verticillata.	Deciduous ditto

HEXANDRIA.

Berberis vulgaris.	Common berberry
———— *v. viol.*	Purple fruited ditto
———— *v. alb.*	White fruited ditto
——— Canadenſis.	Canadian ditto
——— Cretica.	Cretan ditto

OCTAN-

OCTANDRIA.

Koelruteria paniculata.	Panicled Kœlruteria.
Fuchfia coccinea.	Scarlet flower'd Fuchfia
Vaccinium Myrtillus.	Whortleberry
———— pallidum.	Pale ditto
———— ftamineum.	Green wooded ditto
———— uliginofum.	Marfh bilberry
———— diffufum.	Shining leav'd whortleberry
———— anguftifolium.	Narrow leav'd ditto
———— fufcatum.	Cluftered flower'd ditto
———— frondofum.	Obtufe leav'd ditto
———— venuftum.	Red twig'd ditto
———— refinofum.	Clammy ditto
———— amœnum.	Broad leav'd ditto
———— virgatum.	Privet leav'd ditto
———— tenellum.	Gale leav'd dwarf ditto
———— Vitis Idœa.	Red bilberry
———— Oxycoccos.	Common cranberry
———— Macrocarpon.	American ditto
———— arboreum.	Tree whortleberry
———— viride.	Green ditto
———— craffifolium.	Thick leaved ditto
———— Arctoftaphylos.	Madeira ditto
Erica vulgaris.	Common heath
———— v. alb.	White flowered ditto
——— arborea.	Tree ditto
——— tetralix.	Crofs leav'd ditto
———— v. alb.	White flowered ditto
——— Auftralis.	Spanifh ditto
——— herbacea.	Early flowering dwarf ditto
——— multiflora.	Many flowered ditto
——— cinerea.	Fine leav'd ditto
———— v. alb.	White flowered ditto
——— vagans.	
———— v. alb.	
——— Daboecia.	Irifh ditto

Erica Mediterranea.	Mediterranean ditto
Polygonum frutefcens.	Shrubby Polygonum
Daphne. Mezereum.	Common Mezereum
—————— *v. alb.*	White flowered ditto
—————— *v. autum.*	Autumn flowering ditto
——— Tartonraira.	Silvery leaved Daphne
——— Alpina.	Alpine ditto
——— Laureola.	Spurge laurel.
—————— *v. fol. arg. var.*	Silver ftriped ditto
——— Cneorum.	Trailing Daphne
——— Gnidia.	Flax leav'd ditto
Dirca paluftris.	Marfh leatherwood

ENNEANDRIA.

Laurus nobilis.	Common fweet bay
—————— *v. fol. aur. var.*	Gold blotched ditto
——— Benzoin.	Common benjamin tree
——— Saffafras.	Saffafras tree

DECANDRIA.

Sophora tetraptera.	Winged podded Sophora
——— microphilla.	Small leav'd fhrubby ditto
——— Japonica.	Shining leav'd ditto
Anagyris fœtida.	Stinking beantrefoil
Cercis Siliquaftrum.	European Judas tree
——— Canadenfis.	American ditto
Guilandina dioica.	Hardy bonduck
Ruta graveolens.	Common rue
——— montana.	Mountain ditto
——— Calepenfis.	African ditto
Kalmia latifolia.	Broad leav'd Kalmia
——— anguftifolia.	Narrow leav'd red ditto
—————— *v. carnea.*	Narrow leav'd pale ditto
——— Glauca.	Glaucous ditto
Ledum paluftre.	Marfh Ledum
—————— *v. decumb.*	Dwarf ditto

Ledum

Ledum latifolium.	Labrador tea
———— buxifolium.	Box leav'd Ledum
Rhodora Canadenfis.	Canadian Rhodora
Rhododendron ferrugineum.	Rufty Rhododendron
———————— Dauricum.	Daurick ditto
———————— punctatum.	Dotted leav'd ditto
———————— hirfutum.	Hairy ditto
———————— Ponticum.	Purple ditto
——————— *v. fol. var.*	Variegated ditto
———————— maximum.	Rofe coloured ditto
Andromeda Mariana.	Oval leav'd Andromeda
———————— *v. oblong.*	Oblong leav'd ditto
———— ferruginea.	Rufty ditto
——— —— polifolia.	Broad leav'd marfh rofemary
———————— *v. media.*	Common ditto
———————— *v. anguft.*	Narrow leav'd ditto
———————— *v. erect.*	Upright ditto
———— paniculata.	Panicled Andromeda
———— arborea.	Tree ditto
———— racemofa.	Branching ditto
———— axillaris.	Notch'd leav'd ditto
——— —— coriacea.	Thick leav'd ditto
———— acuminata.	Acute leav'd ditto
———— calyculata.	Box leav'd ditto
——— ———— *v. latif.*	Broad leav'd ditto
——— ———— *v. anguft.*	Narrow leav'd ditto
———— ferratifolia.	Serrated ditto
——— —— undulata.	Wave leav'd ditto
Epigea repens.	Creeping Epigea
Gaultheria procumbens.	Trailing Gaultheria
Arbutus Unedo.	Arbutus
———————— *v. rub.*	Scarlet flowered ditto
———————— *v. plen.*	Double flowered ditto
———————— *v. crifp.*	Curled leaved ditto
———— Andrachne.	Oriental ditto

Arbutus

Arbutus Andrachne *v. serratif.* Serrated ditto

———— alpina. Alpine ditto

———— Uva-Urfi. Bearberry ditto

———— thymifolia. Thyme leaved ditto

Clethra alnifolia. Smooth alder leaved Clethra

———————— *v. pubes.* Woolly ditto

———— paniculata. Panicled ditto

Styrax officinale. Officinal Styrax

——— grandifolium. Great leaved ditto

——— lævigatum. Smooth ditto

Hydrangea arborefcens. Shrubby Hydrangea

———————— hortenfis. Changeable flowered ditto

Sedum populifolium. Poplar leaved ftone crop

DODECANDRIA.

Halefia tetraptera. Snowdrop tree

Euphorbia fpinofa. Prickly fpurge

———————— characias. Red ditto

Ariftotelia Macqui. Shining leaved Ariftotelia

————————— *v. fol. aur. var.* Gold blotched ditto.

ICOSANDRIA.

Philadelphus coronarius. Common fyringa

——————— nana. Dwarf ditto

Punica Granatum. Common pomegranate tree

———————————*v. ftam. ftav.* Ditto with yellow ftamina

————————— *v. fl. plen.* Double flowered ditto

Amygdalus Perfica. Common peach tree

———————— *v. glab.* Common nectarine

————————— *v. fl. pleno.* Double flowered peach tree

—————— communis. Sweet almond tree

————————— *v. amara.* Bitter ditto

————————— *v. Chinenfis.* Chinefe ditto

————————— *v. fol. var.* Variegated ditto

————— pumila. Double flow. dwarf almond

————— nana. Common dwarf ditto

Amygdalus

Amygdalus Orientalis.	Silvery leaved ditto
——— Sibirica.	Siberian ditto
Prunus Mahaleb.	Perfumed cherry
——— Armeniaca.	Common apricot
——— nigra.	Black cherry
——— domeſtica.	Common plum
——— inſititia.	Sloe tree
——— pumila.	Dwarf Canadian cherry
Cratægus Aria.	Common white beam tree
——— *v. fol. var.*	Variegated ditto
——— Suecicus.	Swediſh ditto
——— torminalis.	Service tree.
——— coccineus.	Great American hawthorn
——— cordatus.	Maple leaved ditto
——— pyrifolius.	Pear leaved ditto
——— ellipticus.	Oval leaved ditto
——— glanduloſus.	Hollow leaved ditto
——— flavus.	Yellow pear berried ditto
——— parviflorus.	Gooſeberry leaved ditto
——— punctatus.	Great red fruited ditto
——— *v. aur.*	Great yellow fruited ditto
——— prunifolius.	Plum leaved ditto
——— ſanguineus.	Bloody ditto
——— Azarolus.	Parſley leaved ditto
——— Crus-galli.	Common cockſpur ditto
——— *v. pyracanth.*	Pyracantha leaved ditto
——— *v. ſalicif.*	Willow leaved ditto.
——— tanacetifolius.	Tanſey leaved ditto
——— Oxyacanthus.	Common white ditto
——— *v. major.*	Great fruited ditto
——— *v. præcox.*	Glaſtenbury ditto
——— *v. fl. plen.*	Double flowered ditto
——— *v. fl. rub.*	Red flowered ditto
——— *v. fr. flav.*	Yellow berried ditto
——— *v. fr. alb.*	White fruited ditto
——— *v. fol. alb.*	Variegated ditto

Sorbus

Sorbus Aucuparia.	Mountain ash
———————— *v. fol. aur var.*	Gold blotched ditto
——— hybrida.	Baftard fervice tree
——— domeftica.	True fervice ditto
——— Canadenfis.	Canadian ditto
Mefpilus Pyracantha.	Evergreen thorn
———————— *v. diffufa.*	Broad leaved ditto
——— Germanica.	Narrow leaved Dutch medlar
———————— *v. diffufa.*	Broad leaved ditto
——— arbutifolia.	Red fruited arbutus leaved do.
——————— *v. nigr.*	Black fruited ditto
——————— *v. alb.*	White fruited ditto
——————— *v. latif.*	Broad leaved ditto
——— Amelanchier.	Alpine ditto
——————— *v. pend.*	Weeping ditto
——— Chamœ-Mefpilus.	Dwarf quince
——— Canadenfis.	Snowy Mefpilus
——— cotoneafter.	Dwarf ditto
——— tomentofa.	Quince leaved ditto
——————— *v fl. plen.*	Double flowered ditto
——————— *v. fol. arg. var.*	Gold blotched ditto
——— latifolia.	Broad leaved ditto
——— Japonica.	Japan Mefpilus
Pyrus communis.	Common pear tree
——————— *v. fl. plen.*	Double flowered ditto
——————— *v. fol. varieg.*	Variegated ditto
——— Pollveria.	Woolly leaved ditto
——— Malus	Common apple tree
——————— *v. fol. varieg.*	Variegated ditto
——————— *v. Sibirica.*	Siberian crab
——— fpectabilis.	Chinefe apple
——— baccata.	Small fruited crab
——— coronaria.	Sweet fcented ditto
——— anguftifolia.	Narrow leaved ditto
——— Cydonia.	Quince tree
——— falicifolia.	Willow leaved crab
——— præcox.	Early ditto

Spiræa

Spiræa lævigata.	Smooth leaved Spiræa
—— falicifolia *v. carn.*	Flefh coloured willow leaved do.
—————— *v. panicul.*	Panicled ditto
—————— *v. latif.*	Broad leaved ditto
—————— *v. alb.*	White flowered ditto
—— tomentofa.	Scarlet ditto
—— hypericifolia.	Hypericum Frutex
—— crenata.	Hawthorn leaved Spiræa
—— opulifolia.	Virginian gilder rofe
—— forbifolia.	Service-tree leaved Spiræa
—— thalictroides.	Meadow-rue leaved ditto
Rofa lutea *v. fimp.*	Single yellow rofe
—————— *v. bicolor.*	Red and yellow Auftrian ditto
—— fulphurea.	Double yellow ditto
—— blanda.	Hudfon's bay ditto
—— cinnamomea *v. fl. fimp.*	Single cinnamon ditto
—————— *v. fl. plen.*	Double ditto
—— arvenfis.	White dog ditto
—— pimpinellifolia.	Small burnet leaved ditto
——fpinofiffima *v. commun.*	Common Scotch ditto
—————— *v. fl. rub.*	Red Scotch ditto
—————— *v. fl. varieg.*	Striped flowered ditto
—————— *v. fl. plen.*	Double Scotch ditto
—————— *v. fpiral.*	Spiral ditto
—— Carolinienfis *v. max.*	Great fingle burnet leaved ditto
—————— *v. plen. max.*	Great double ditto
—————— *v. fimp.*	Penfylvanian ditto
—————— *v. pleno.*	Double ditto
—————— *v. patula.*	Spreading ditto
—————— *v. erect.*	Upright ditto
—— villofa *v. fimp.*	Single apple ditto
—————— *v. plen.*	Double ditto
—— Provincialis *v. commun.*	Common Provence ditto
—————— *v.*	Childings ditto
—————— *v. fl. rub.*	Red ditto
—————— *v. pallida.*	Blufh ditto

Rofa

Rosa Provincialis *v. fl. alb.*	White ditto
———————— *v. fl. plen.*	Double white ditto
———— *v. de Meux major.*	Great dwarf rose de meux
———— *v. de Meux min.*	Small dwarf ditto
—— centifolia *v.*	Dutch 100 leaved rose
———————— *v.*	Blush ditto
———————— *v.*	Singleton's ditto
———————— *v.*	Petite ditto
———————— *v.*	Burgundy rose
———————— *v.*	Single velvet ditto
———————— *v.*	Double velvet ditto
———————— *v.*	Sultan ditto
———————— *v.*	Stepney ditto
———————— *v.*	Garnet ditto
———————— *v.*	Bishop ditto
———————— *v.*	Lisbon ditto
—— Gallica. *v. officinal.*	Red officinal ditto
———————— *v.*	Mundi rose
———————— *v.*	Marbled ditto
———————— *v.*	Virginian ditto
—— Damascena *v.*	Red damask ditto
———————— *v.*	Blush ditto
———————— *v.*	York and Lancaster ditto
———————— *v.*	White monthly ditto
———————— *v.*	Blush monthly ditto
———————— *v.*	Blush Belgick ditto
———————— *v.*	Red Belgick ditto
———————— *v.*	Great royal ditto
—— sempervirens.	Evergreen ditto
—— semperflorens.	Ever flowering ditto
—— Indica.	Indian ditto
—— pumila.	Dwarf Austrian ditto
—— turbinata.	Frankfort ditto
—— rubiginosa *v.*	Common sweet ditto
———————— *v.*	Semidouble ditto
———————— *v.*	Double ditto

Rosa

Rofa rubiginofa *v.*	Moffy double ditto
——————— *v.*	Evergreen double ditto
——————— *v.*	Marbled double ditto
——————— *v.*	Rofe double ditto
—— mufcofa.	Mofs Provence rofe
—— mofchata *v.*	Single mufk ditto
——————— *v.*	Double mufk ditto
——————— *v.*	Red mufk ditto
—— alpina *v.*	Alpine ditto
——————— *v.*	Red Alpine ditto
—— canina *v.*	Dog rofe
——————— *v.*	Variegated ditto
—— pendulina.	Smooth pendulous ditto
—— alba.	Single white rofe
——————— *v.*	Double ditto
——————— *v.*	Small maidens blufh ditto
——————— *v.*	Great ditto
——————— *v.*	Clustered ditto
——————— *v.*	Small burnet ditto
Rofa.	Portland ditto
——	Labrador ditto
——	Stœban ditto
——	Penfylvanian ditto
——	De Rheims
——	Blandford rofe
——	Thornlefs ditto
——	American upright ditto
——	Pluto ditto
——	Hedgehog ditto
——	Atlas ditto
——	Giant ditto
——	Bright purple ditto
——	Plicate ditto
——	Caucafia
——	Pyramidal ditto
——	Cardinal ditto

E   Rubus

Rubus Idæus	Raſpberry
———— v. alb.	White ditto
———— v. lævigat.	Smooth ditto
—— Occidentalis.	Virginian ditto
—— hiſpidus.	Briſtly bramble
—— Cœſius.	Dewberry ditto
—— fruticoſus.	Common ditto
———— v. alb.	White fruited ditto
———— v. pleno.	Double flowered ditto
———— v. inermis.	Smooth ditto
———— v. laciniat.	Parſley leaved ditto
—— villoſus.	Hairy bramble
—— ſaxatilis.	Stone ditto
—— Arcticus.	Dwarf bramble
Potentilla fruticoſa.	Shrubby cinquefoil
Calycanthus floridus.	Carolina allſpice
———— v. ovatus.	Round leaved ditto

POLYANDRIA.

Thea Bohea.	Broad leaved tea
———— v. ſtrict.	Narrow leaved ditto
Gordonia pubeſcens.	Loblolly bay
Ciſtus populifolius.	Great poplar leaved Ciſtus
———— v. minor.	Small ditto
—— laurifolius.	Laurel leaved ditto
—— ladaniferus.	Common gum ditto
———— v. planifolius.	Flat leaved ditto
———— v. fl. rub.	Red flowered ditto
———— v. fol. aur. var.	Gold blotched ditto
—— Monſpelienſis.	Montpellier ditto
—— laxus.	Wave leaved ditto
—— ſalvifolius.	Sage leaved ditto.
—— incanus.	Hoary leaved ditto
—— albidus.	White leaved ditto
—— Creticus.	Cretan ditto
—— criſpus.	Curled leaved ditto

Ciſtus

Ciſtus criſpus *v. latif.*	Broad leaved ditto
—— halimifolius.	Sea purſlane leaved ditto
——————— *v. oleæfol.*	Olive leaved ditto
—— umbellatus.	Umbelled ditto
—— ſcabroſus.	Rough ditto
—— undulatus.	Undulated ditto
—— obliquus.	Olive ditto
—— canus.	Myrtle leaved ditto
—— marifolius.	Marum leaved Ciſtus
—— Anglicus.	Hairy ditto
—— Surrejanus.	Small flowered ditto
—— ſerpyllifolius.	Serpyllum leaved ditto
—— thymifolius.	Thyme leaved ditto
—— Helianthemum	Yellow flowered dwarf- ditto
——————— *v. ſulphur.*	Sulphur ditto
——————— *v. alb.*	White ditto
——————— *v. roſeo.*	Roſe ditto
——————— *v. purp.*	Purple ditto
——————— *v. fol. aur. var.*	Gold blotched ditto
—— Apenninus.	Apennine ditto
—— polifolius.	Mountain ditto
Fothergilla alnifolia.	Broad leaved Fothergilla
——————— *v. acuta.*	Narrow leaved ditto
——— ſpecioſa	Snowy ditto
——— glauca	Glaucous ditto
Calligonum Pallaſia	
Annona triloba.	Trifid fruited cuſtard apple
Atragene alpina.	Alpine Atragene
Clematis cirrhoſa.	Evergreen virgins bower
——————— *v. fl pleno.*	Double flowered ditto
——— Viticella.	Purple virgins bower
——————— *v. fl. pleno.*	Double flowered ditto
——————— *v. fl. rub.*	Red flowered ditto
—— Viorna.	Leathery flowered ditto
—— criſpa,	Curled ditto
—— Orientalis.	Oriental ditto

Clematis

Clematis Virginiana.	Virginian ditto
——— Vitalba.	Travellers Joy
——— Apennina.	Apennine ditto
——— flamula.	Sweet scented virgins bower

DIDYNAMIA.

Hyssopus officinalis	Common hyssop
——————— *v. fl. rub.*	Red flowered ditto
——————— *v. fl. alb.*	White flowered ditto
——————— *v. fol. pilof.*	Hairy ditto
Teucrium flavum *v. fol.* ⎫ *crenat.* ⎭	Notched leaved yellow flowered shrubby germander
——————*v. fol. integris.*	Entire leaved ditto
——— montanum.	Dwarf germander
——— Polium *v. mont.*	White mountain ditto
——— *v. marit.*	Sea ditto
——— capitatum.	Round headed ditto
Satureja montana.	Winter savory
——— capitata.	Ciliated ditto
Lavandula spica.	Common lavender
——————— *v. fl. alb.*	White flowered ditto
——————— *v. latifol.*	Broad leaved ditto
Sideritis incana.	Lavender leaved iron wort
Stachys latifolia.	Broad leaved Stachys
Phlomis fruticofa *v. anguft.*	Narrow leaved shrubby Phlomis
——————— *v. latif.*	Broad leaved ditto
———purpurea.	Purple ditto
Thymus Serpyllum	Common smooth mother of
——————— *v. citri odore.*	Lemon thyme [thyme
——————— *v. villofus*	Hoary mother of thyme
——————— *v. hirtus.*	Hairy ditto
———vulgaris.	Narrow leaved garden ditto
——————— *v. fol. arg. var.*	Silver blotched garden thyme
——————— *v. latiore.*	Broad leaved ditto
——— Zygis.	Linear leaved ditto

Bignonia

Bignonia Catalpa.	Common Catalpa
—— capreolata.	Four leaved trumpet flower
—— radicans.	Great afh leaved ditto
———————— *v. minor*.	Small ditto
Linnæa Borcalis.	Two flowered Linnæa
Vitex Agnus Caft. *anguft*.	Narrow leaved chafte tree
———————— *v. latifol*.	Broad leaved ditto

TETRADYNAMIA.

Iberis fempervirens.	Nar. leav. evergreen candy-tuft
Alyffum faxatile.	Shrubby madworth
Vella Pfeudo-Cytifus.	Shrubby Vella

MONADELPHIA.

Hibifcus fyriacus.	Althæa Frutex
———— *v. rub*.	Red ditto
———— *v. alb*.	White ditto
———— *v. variegat*.	Striped flowered ditto
Stuartia Malacodendron	Common Stuartia
—— Marylandica.	Maryland ditto

DIADELPHIA.

Polygala Chamœbuxus.	Box leaved milkwort
Spartium junceum.	Common Spanifh broom
———————— *v. pleno*.	Doubled flowered ditto
—— decumbens.	Trailing ditto
—— fcorpius.	Scorpion ditto
—— multiflorum.	Portugal white ditto
—— fcoparium.	Common ditto
—— patens.	Woolly podded ditto
—— radiatum.	Starry ditto
Genifta candicans.	Montpellier cytifus
—— triquetra.	Triangular broom
—— fagittalis.	Jointed ditto
—— tinctoria.	Common dyer's ditto

Genifta

Genista florida.	Spanish ditto
———— pilosa.	Hairy ditto
———— Anglica.	Petty whin
———— Germanica.	German broom
———— Hispanica.	Dwarf prickly ditto
———— Lusitanica.	Portugal ditto
Ulex Europeus.	Common furze
———— nanus.	Dwarf ditto
Amorpha fruticosa.	Shrubby bastard indigo
Ononis fruticosa.	Shrubby restharrow
———— rotundifolia.	Round leaved ditto
Glycine frutescens.	Common kidney bean tree
Cytisus Laburnum.	Common Laburnum
———————— *v. fol. aur. var.*	Gold blotched ditto
———————— *v. alpinus.*	Alpine ditto
———— nigricans.	Black Cytisus
———— divaricatus.	Clammy ditto
———— Wolgaricus.	Winged leaved ditto
———— hirsutus.	Hairy ditto
———— sessilifolius.	Common ditto
———— capitatus.	Clustered flowered ditto
———— purpureus.	Purple ditto
———— Austriacus.	Siberian ditto
———— supinus.	Trailing ditto
———— argenteus.	Silvery ditto
———— biflorus.	Smooth ditto
Robinia Psued-Acacia.	Two thorned acacia
———— hispida.	Rose ditto
———— Caragana.	
———— spinosa.	Thorny Robinia
———— Altagana.	
———— Chamlagu.	Shining ditto
———— Halodendron.	Salt tree ditto
———— frutescens.	Shrubby ditto
———— pygmæa.	Dwarf ditto
Colutea arborescens.	Common bladder senna

Colutea

Colutea cruenta.	Oriental ditto
——— Pocockii.	Pocock's ditto
Coronilla emerus.	Scorpion senna
——— Vallentina.	Small shrubby Coronilla
——— Glauca.	Great shrubby ditto
Lotus Dorycnium.	Shrubby birds foot trefoil
——— hirsutus.	Hairy ditto
Astragalus Tragacantha.	Goats thorn milk vetch
Medicago arborea.	Moon trefoil, or tree medick

POLYADELPHIA.

Hypericum calycinum.	Great flowered St. John's wort
——— Androsœmum.	Common Eutsan
——— Olympicum.	Olimpian St. John's wort
——— elatum.	Tall ditto
——— hircinum.	Common stinking shrubby ditto
——— v. minus.	Small ditto
——— prolificum.	Proliferous ditto

SYNGENESIA.

Santolina Chamæ-Cyparis.	Common lavender cotton
——— tomentosa.	Hoary ditto
——— rosmarinifolia.	Rosemary leaved ditto
Artemisia Abrotanum.	Common southernwood
——— Santonica.	Tartarian ditto
——— crithmifolia.	Samphire leaved ditto
Baccharis halimifolia.	Virginian groundsel tree
Othonna cheirifolia.	Stock leaved African ragworth

MONOECIA.

Passiflora cærulea.	Common passion flower
Aristolochia Sipho.	Broad leaved birtworth
——— arborea.	Tree ditto
——— sempervirens.	Evergreen ditto

Axyris

Axyris ceratoides.	Shrubby Axyris
Comptonia afplenifolia.	Fern leaved Comptonia
Buxus fempervirens.	Common tree box
——————— *v. lanceolat.*	Narrow leaved ditto
——————— *v. fol. aur.*	Gold blotched ditto
——————— *v. fol. argent.*	Silver blotched ditto
——————— *v. fol. apice notat.*	Horfe fhoe ditto
—— humilis.	Dwarf ditto
—— Balearica.	Minorcan ditto
Iva frutefcens.	Baftard Jefuit's bark tree
Corylus roftrata.	American cuckold nut
—— Colurna.	Conftantinople hazel nut
—— Avellana.	Common ditto
——————— *v. alb.*	White filbert
——————— *v. rubra.*	Red ditto
——————— *v. grandis.*	Cob nut tree
——————— *v. glomerata.*	Cluftered nut ditto
Liquidambar ftyraciflua	Sweet gum
——————— imberbe.	Oriental Liquidamber
Thuja Occidentalis.	American arbor vitæ tree
—— Orientalis.	China ditto
Cupreffus fempervirens.	Upright cyprefs tree
——————— *v. horizontalis.*	Male fpreading ditto
——————— difticha.	Common deciduous cyprefs
——————— *v. nutan.*	Long leaved ditto
——————— thyoides.	White cedar
——————— pendula.	Portugal cyprefs
——————— juniperoides.	Juniper leaved ditto
Croton febiferum.	Tallow tree

DIOECIA.

Salix hermaphrodita.	Shining willow
—— triandra.	Smooth ditto
—— pentandra.	Sweet ditto
—— vitellina.	Yellow ditto

Salix

Salix amygdalina.	Almond leaved ditto
—— haftata.	Halbert leaved ditto
—— fragilis.	Crack ditto
—— Babylonica.	Weeping ditto
—— rubra.	Red ditto
—— glauca.	Glaucous, or alpine ditto
—— myrfinitis.	Whorl leaved ditto
—— herbacea.	Herbaceous ditto
—— retufa.	Blunt leaved ditto
—— reticulata.	Wrinkled ditto
—— myrtilloides.	Myrtle leaved ditto
—— aurita.	Small round leaved ditto
—— lanata.	Downy ditto
—— Lapponum.	Lapland ditto
—— arenaria.	Sand ditto
—— incubacea.	Trailing ditto
—— repens.	Creeping ditto
—— fufca.	Brown ditto
—— rofmarinifolia.	Rofemary leaved ditto
—— capræa.	Common ditto
—— cinerea.	Afh coloured ditto
—— triftis.	Narrow leaved American ditto
—— viminalis.	Ofier ditto
—— alba.	White ditto
—— variegata.	Variegated ditto
—— Septentrionalis.	American ditto
—— Penfylvanica.	Penfylvanian ditto
—— acuminata.	Sharp pointed leaved ditto
—— fiffa.	Cloven leaved ditto
—— Japonica.	Japan ditto
Empetrum nigrum.	Crow berries
——— Caledonum.	Scotch ditto
Vifcum album.	Common miffeltoe
Hippophae Rhamnoides.	Sea buckthorn
——— Canadenfis.	Canadian ditto

F   Myrica

Myrica Gale.	Sweet gale [myrtle
—— cerifera *v. angust.*	Common American candleberry
——————— *v. latif.*	Broad leaved ditto
—— Faya.	Azorian ditto
Zanthoxylon Clava-Herculis.	Common toothache tree
Pistacia officinalis.	Pistachia tree
—— Terebinthus.	Common turpentine tree
Smilax aspera.	Rough bindweed
——————— *v. auriculata.*	Ear leaved ditto
—— Sarsaparilla.	Sarsaparilla
—— rotundifolia.	Round leaved smilax
—— laurifolia.	Laurel leaved ditto
—— tamnoides.	Black briony leaved ditto
—— caduca.	Deciduous ditto
—— Bona-nox.	Ciliated ditto
—— lanceolata.	Spear leaved ditto
Mercurialis tomentosa.	Woolly mercury
Coriaria myrtifolia.	Myrtle leaved sumach
Menispermum Canadense.	Canadian moonseed
——————Carolinum.	Carolina ditto
Juniperus thurifera.	Spanish juniper
—— Bermudiana.	Bermudas cedar
—— Sabina.	Common savin
——————— *v. tamariscifol.*	Tamarisk leaved ditto
——————— *v. fol. varieg.*	Variegated ditto
—— Virginiana.	Virginian red cedar
—— communis.	Common juniper
——————— *v. montana.*	Procumbent ditto
——————— *v. Suecica.*	Swedish ditto
—— Oxycedra.	Brown berried ditto
—— Phœnicea.	Phœnician ditto
Ephedra distachya.	Great shrubby horsetail
—— monostachya.	Small ditto
Cissampelos smilacina.	Smilax leaved Cissampelos
Ruscus aculeatus.	Prickly butchers broom
—— laxus.	

Ruscus

Rufcus hypophylus.   Broad leaved ditto
———— hypogloffus.   Double leaved ditto
———— racemofus.   Alexandrian laurel

POLYGAMIA.

Atriplex Halimus   Spanifh fea purflane
———— portulacoides.   Common ditto
Celtis Auftralis.   European nettle tree
—— Orientalis.   Oriental ditto
—— Occidentalis.   American ditto
Ailanthus glandulofa.   Tall Ailanthus
Gleditfia triacanthos.   Three thorned acacia
————— *v. monofper.*   Single feeded, or water ditto
———— horrida.   Strong fpined ditto
Diofpiros Lotus.   European date plum
———— Virginiana.   American ditto
Nyffa integrifolia.   Mountain tupelo
—— denticulata.   Water ditto
Ficus Carica.   Common fig tree

No. 14. & 15.

Containing the GRAMINA VERA *or* TRUE GRASSES, *including Oats, Wheat, Rye, and Barley.*

☞ NOTE.—The Marks are in the form of BATTLE DORES, BLACK and RED Letters on a WHITE GROUND, and the arrangement begins alphabetically, at the East end of the Garden.

Ægilops ovatus.	Oval spiked hard grass
——— triuncialis.	Long spiked ditto
Agrostis alba.	White marsh bent grass
——— alpina.	Alpine ditto
——— arenaria.	Sea ditto
——— canina.	Brown ditto
——— Cornucopiæ.	Cornucopiæ ditto
——— Mexicana.	Mexican ditto
——— miliacea.	Millet ditto
——— minima.	Very small ditto
——— palustris.	Marsh ditto
——— pumila.	Dwarf ditto
——— repens.	Creeping ditto
——— rubra.	Red ditto
——— setacea.	Bristly ditto
——— Spica-Venta.	Silky ditto
——— sylvatica.	Wood ditto
——— tenuifolia.	Slender leaved ditto
——— verticillata.	Whorled ditto
——— vinealis.	Vine ditto
Aira aquatica.	Water hair grass

Aira

Aira Canefcens.	Gray ditto
—— caryophyllea.	Silvery leaved ditto
—— cæfpitofa.	Turfy ditto
—— flexuofa.	Heath ditto
—— montana.	Mountain ditto
—— præcox.	Early ditto
Alopecurus agreftis.	Field foxtail grafs
—— bulbofus	Bulbous ditto
—— geniculatus.	Flote ditto
—— Monfpelienfis.	Bearded ditto
—— paniceus.	Small bearded ditto
—— pratenfis.	Meadow ditto
Anthoxanthum odoratum.	Sweet fcented fpring grafs
Arundo arenaria.	Sea reed grafs
—— Calamagroftis.	Wood ditto
—— Donax.	Manured ditto
—— Fpigejos.	Small ditto
—— Phragmites.	Common ditto
Avena elatior.	Tall oat grafs
—— var. nodofa.	Bulbous rooted ditto
—— var. muticis.	Tall oat grafs without awns.
—— fatua.	Wild ditto
—— flavefcens.	Yellow ditto
—— fragilis.	Brittle ditto
—— nuda.	Naked ditto
—— Penfylvanica.	Penfylvanian ditto
—— pratenfis.	Meadow ditto
—— pubefcens.	Soft ditto
—— fativa nigra.	Cultivated black, or common do.
—— fativa alba.	Cultivated white, or common do.
—— Sibirica.	Siberian ditto
—— fterilis.	Bearded ditto
—— ftrigofa.	Scrannel ditto
Briza maxima.	Great quaking grafs
—— media.	Middle ditto
—— minor.	Small ditto

Briza

Briza virens.	Spanish ditto
Bromus arvensis.	Corn brome grass
——— ciliatus.	Ciliated ditto
——— erectus.	Upright ditto
——— giganteus.	Tall ditto
——— gracilis	Slender ditto
——— hirsutus.	Wood brome grass.
——— inermis.	Smooth ditto
——— Madritensis.	Wall ditto
——— mollis.	Soft ditto
——— secalinus.	Field ditto
——— squarosus.	Corn ditto
——— sterilis.	Barren ditto
——— stipoides.	Feathery ditto
——— sylvaticus.	Hairy stalked ditto
——— tectorum.	Nodding panicled ditto
Cynosurus cæruleus.	Blue dog's-tail grass
——— cristatus.	Crested ditto
——— echinatus.	Rough ditto
——— erucæformis.	Linear spiked ditto
Dactylis Cynosuroides.	American cock's-foot grass
——— glomerata.	Rough ditto
——— patens.	Spreading ditto
——— stricta.	Sea ditto
Elymus arenarius.	Sea lyme grass
——— Canadensis.	Canadian ditto
——— caninus.	Dog's ditto
——— Europeus.	Wood ditto
——— geniculatus.	Jointed ditto
——— giganteus.	Gigantic ditto
——— Hystrix.	Rough ditto
——— Philadelphicus.	Philadelphian ditto
——— Sibiricus.	Siberian ditto
——— Virginicus.	Virginian ditto
Festuca bromoides.	Barren Fescue grass
——— Cambrica.	Welsh ditto

Festuca

Feſtuca decumbens.	Decumbent ditto
——— duriuſcula.	Hard ditto
——— dumetorum.	Pubeſcent feſcue graſs.
——— elatior.	Tall ditto
——— fluitans.	Flote ditto
——— glauca.	Glaucous ditto
——— loliacea.	Darnel ditto
——— ovina.	Sheep ditto
——— Var. 2.	Awned ſheeps ditto
——— Var. 3.	Awnleſs ſheeps ditto
——— Vvr. 4.	Viviparous ſheeps ditto
——— pratenſis.	Meadow ditto
——— rubra.	Red ditto
——— ſpadicea.	Spadiceous ditto
——— unigluma.	Sea ditto
Holcus lanatus.	Meadow ſoft graſs
——— mollis.	Creeping ditto
——— odoratus.	Sweet ſcented ditto
Hordeum diſtichon.	Common barley
——— hexaſtichon.	Winter ditto
——— jubatum.	Long bearded ditto
——— maritinum.	Sea ditto
——— marinum.	Wall ditto
——— pratenſe.	Meadow ditto
——— ſylvaticum.	Wood ditto
——— vulgare.	Spring ditto
——— Zeocriton.	Battledore ditto
Lagurus ovatus.	Oval ſpiked Lagurus
Lolium arvenſe.	White darnel graſs
——— perenne	Ray ditto
——— temulentum.	Rivery
——— v. nud.	naked ditto
——— tenue.	Slender darnel graſs
Melica altiſſima.	Tall melic graſs
——— cærulea.	Purple ditto
——— ciliata.	Ciliated ditto

Melica

Melica nutans.	Mountain ditto
———— fericea.	Silky ditto
———— uniflora.	Single flowered wood ditto
Milium effufum.	Common millet grafs
———— Lendigerum.	Yellow fpiked ditto
———— paradoxum.	Black feeded ditto
Nardus ftricta.	Matt grafs
Panicum capillare.	Hair panicled panic grafs
———— Crus-galli.	Thick fpiked cocks foot ditto
———— Dactylon.	Creeping ditto
———— filiforme.	Clofe fpiked ditto
———— glaucum.	Glaucous ditto
———— Italicum.	Italian ditto
———— miliaceum.	Millet ditto
———— fanguinale.	Slender fpiked ditto
———— verticillatum.	Rough ditto
———— viride.	Green ditto
Phalaris aquatica.	Water canary grafs
———— arundinacea.	Reed ditto
———————————— *v. f. arg.*	Silver ftriped ditto
———————————— *v. f. aur.*	Gold ftriped reed ditto
————Canarienfis.	Manured ditto
———— paradoxa.	Briftly fpiked ditto
———— phleoides.	Cat's-tail ditto
Phleum alpinum.	Mountain Timothy grafs
———— arenarium.	Sea ditto
———— nodofum.	Bulbous cat's-tail grafs
———— paniculatum.	Branched Timothy grafs
———— pratenfe.	Common ditto
Poa alpina.	Alpine meadow grafs
———— anguftifolia.	Narrow leaved ditto
———— annua.	Annual ditto
———— aquatiea.	Water ditto
———— comprefsa.	Creeping ditto
———— criftata.	Crefted ditto
———— diftans.	Loofe flowered ditto

Poa glauca.	Glaucous ditto
—— loliacea.	Spiked sea ditto
—— maritima.	Sea ditto
—— nemoralis.	Wood ditto
—— pratenfis.	Great ditto
—— retroflexa.	Retroflexed ditto
—— rigida.	Hard ditto
—— trivialis.	Rough ftalked ditto
Rottbœllia incurvata.	Sea hard ditto
Secale cereale *hybernum.*	Winter manured rye
——————— *vernum.*	Spring ditto
Triticum æftivum.	Summer wheat
—— compofitum..	Compound ditto
—— hybernum.	Winter, or Lamas ditto
—— junceum.	Rufh wheat grafs
—— maritimum.	Sea ditto
—— monococcum.	One grained wheat
—— Polonicum.	Poland ditto
—— repens.	Couch grafs
—— Spelta.	Spelt wheat
—— tenellum.	Dwarf wheat grafs
—— turgidum.	Cone wheat
——————— *var. ariftis nigris. with black awns.*	

No. 28.

The HORTUS TINCTORIUS, *or* DYER's DIVISION, *containing such Plants as are subservient to the Purposes of Dyeing.*

☞ NOTE.—The Marks are in the Form of small CROSSES, BLACK and RED Letters on a FRENCH WHITE GROUND, towards the West end of the Garden.

Acanthus mollis.	Smooth bears breech
Ajuga reptans.	Common bugle
Anchusa officinalis.	Officinal buglofs
Anemone Pulsatilla.	Pasque-flower Anemone
Anthemis tinctoria.	Yellow camomile
Anthyllis vulneraria.	Yellow kidney vetch
Arbutus Uva-Ursi.	Trailing Arbutus
Arundo Phragmites.	Common reed grafs
Berberis vulgaris.	Common berberry
Betula alba.	Common birch tree
—— Alnus.	Common alder tree
—— nana.	Smooth dwarf birch
Bidens tripartita.	Water hemp agrimony
Bixa Orellana.	Heart-leaved Bixa
Bromus fecalinus.	Field brome grafs
Cæfalpinia Sappan.	Narrow-leaved prickly Brafiletto
Caltha paluftris.	March marygold
Campanula rotundifolia.	Round-leaved bell flower
Carpinus Betulus.	Common hornbeam
Centaurea Cyanus.	Blue bottle
Centaurea Jacea.	Common Centaury
Chærophyllum fylveftre.	Cow weed
Comarum paluftre.	Marfh cinquefoil

Crocus

Crocus fativus.	Saffron
Datifca canabina.	Baftard hemp
———— hirta.	Hairy Datifça
Delphinium Confolida.	Wild lark fpur
Erica vulgaris.	Common heath
Fraxinus excelfior.	Afh tree
Galium verum.	Yellow ladies bedftraw
Genifta tinctoria.	Dyers broom
Hieracium umbellatum.	Umbelled hawk-weed
Hypericum perfoliatum.	Perfoliate St. John's wort.
Indigofera tinctoria.	Dyer's indigo
Iris Germanica	German Iris
Iris Pfeud-acorus.	Yellow Iris, or Common flagger
Ifatis tinctoria.	Common dyer's woad
Lawfonia inermis.	Smooth Lawfonia
Lichen calcareus.	Dyer's Lichen
———— caperatus.	Rofe ditto
———— juniperinus.	Juniper ditto
———— omphalodes.	Purple ditto
———— pulmonarius.	Lungwort ditto
———— puftulatus.	Singed ditto
———— faxatilis.	Stone ditto
———— vulpinus.	Fox ditto
Liguftrum vulgare.	Common privet
Lithofpermum arvenfe.	Baftard alkanet
Lycopodium complanatum.	Smooth club mofs
Lycopus Europeus.	Water horehound
Lyfimachia vulgaris.	Common loofe-ftrife
Myrica Gale.	Sweet Gale
Nymphæa alba.	White water lilly
Polygonum Perficaria.	Spotted arfmart
Potentilla Tormentilla.	Tormentil
Pyrus communis.	Pear tree
——— Malus.	Common apple tree
Quercus Ægilops.	Great prickly cuped oak
———— Robur.	Common oak tree

Refeda

Reseda Luteola.	Dyer's weed
Rhamnus catharticus.	Purging buckthorn
———— frangula.	Berry bearing alder
———— minor.	Dyer's grains
Rhus Coriaria.	Elm-leaved Sumack
——— Colinus.	Venus's sumack
Rosa spinosissima.	Common Scotch rose
Rubia tinctorum.	Dyer's madder
Rubus fruticosus.	Common bramble, or blackberry
Rumex Acetosa.	Common sorrel
———— acutus.	Sharp pointed dock
———— maritimus.	Golden dock
Salix pentandra.	Sweet willow
Sambucus Ebulus.	Dwarf elder
———— nigra.	Common ditto
Satyrium nigrum.	Black-flowered Satyrion
Scabiosa succisa.	Devil's-bit scabious
Senecio Jacobœa.	Common ragwort
Serratula tinctoria.	Common saw-wort
Stachys sylvatica.	Hedge nettle
Tanacetum vulgare.	Common tansey
Thalictrum flavum.	Common meadow rue
Thapsia villosa.	Deadly carrot
Tormentilla erecta.	Common tormentil
Urtica dioica.	Common nettle
Viola odorata.	Sweet violet
Xanthium Strumarium.	Lesser burdock

CATALOGUE

OF

No. XIX.

IN

THE MAP OF THE

DUBLIN SOCIETY'S

BOTANIC GARDEN, AT GLASNEVIN.

CONTAINING THE

HERBARIUM,

OR

SYSTEMATIC ARRANGEMENT

OF

HERBACEOUS PLANTS,

𝕮𝖗𝖞𝖕𝖙𝖔𝖌𝖆𝖒𝖎𝖈𝖐𝖘 𝕰𝖝𝖈𝖑𝖚𝖉𝖊𝖉.

☞ NOTE. At the beginning of each CLASS, a Mark is fixed, expreſſive of the Claſs in the Linnæan Syſtem, about Eighteen Inches high; and the Marks of the *Orders, Genera, Species, and Varieties,* are in the Form of ſmall Croſſes, with BLACK and RED LETTERS on a WHITE GROUND, which begin at the Front of the Houſe, and run on to the W. End of the Garden.

HERBARIUM,

OR

HERBACEOUS DIVISION.

CLASSIS I.

MONANDRIA MONOGYNIA

Salicornia herbacea.	Marſh jointed glaſſwort
Hippuris vulgaris.	Mares-tail

MONANDRIA DIGYNIA.

Coriſpermum hyſſopifolium.	Hyſſop leaved tickſeed
Callitriche verna	Vernal ſtarheaded chickweed
———— autumnalis.	Autumnal ditto
Blitum capitatum.	Berry-headed ſtrawberry blite
——— virgatum.	Slender branched ditto
——— *v. chenopodioides.*	Gooſe-foot leaved ditto
Cinna arundinacea	

CLASSIS II.

DIANDRIA MONGYYNIA.

Circæa Lutetiana.	Enchanters nightſhade.
——— alpina.	Mountain ditto
Veronica Sibirica.	Siberian ſpeedwell
——— Virginica.	Virginian white flowered ditto
——————— *v. incarnata.*	Virginian bluſh ditto
——— ſpuria.	Baſtard ditto
——— maritima.	Blue flowered ſea ditto
——— *v. fl. alb.*	White flowered ditto
——— *v. fl. incarnata.*	Fleſh coloured ditto

Veronica

Veronica longifolia.	Long leaved ditto
—— incana.	Hoary ditto
—— candida.	White ditto
—— spicata.	Upright spiked male ditto
—— hybrida.	Welsh ditto
—— pinnata.	Winged leaved ditto
—— laciniata.	Jagged-leaved ditto
—— incisa.	Cut-leaved ditto
—— officinalis.	Officinal male ditto
—— Allionii.	
—— aphylla.	Naked stalked ditto
—— bellidioides.	Daisy-leaved ditto
—— fruticulosa.	Purple stalked evergreen ditto
—— saxatilis.	Rock ditto
—— alpina.	Alpine ditto
—— serpyllifolia.	Smooth ditto
—— tenella.	Moneywort leaved ditto
—— Beccabunga.	Brooklime
—— Anagallis	Longleaved water speedwell
—— scutellata.	Narrow-leaved water ditto
—— Teucrium.	Hungarian ditto
—— prostata.	Trailing ditto
—— montana.	Mountain ditto
—— chamædrys.	Wild Germander, or speedwell
—— Orientalis.	Oriental ditto
—— multifida.	Multified ditto
—— Austriaca.	Austrian ditto
—— urticæfolia.	Nettle-leaved ditto
—— latifolia.	Broad leaved ditto
—— paniculata.	Panicled ditto
—— agrestis.	Blue flowered germander ditto
—————— v. fl. alb.	White flowered ditto
—— hederifolia.	Blue flowered ivy-leaved ditto
—————— v. fl. alb.	White flowered ditto
—— triphyllos.	Trifid ditto
—— verna.	Spring ditto

Veronica

Veronica peregrina.	Knot-grafs leaved ditto
Gratiola officinalis.	Hedge hyffop
Pinguicula Lufitanica.	Hairy ftalked butterwort
———— vulgaris.	Common butterwort
Utricularia vulgaris.	Common hooded water milfoil
———— minor.	Leffer ditto
Verbena haftata.	Halbert leaved vervain
——— urticæfolia.	Nettle leaved ditto
——— officinalis.	Officinal ditto
——— fupina.	Trailing ditto
Lycopus Europeus.	Water horehound
——— exaltatus.	Tall Lycopus
——— Virginicus.	Virginian Lycopus
Amethyftea cærulea	Blue amethyft.
Cunila pulegioides,	Penny royal-leaved Cunila
——— thymoides.	Thyme leaved ditto
Ziziphora capitata.	Oval leaved Ziziphora
——— tenuior.	Spear leaved ditto
Monarda fiftulofa.	Purple Monarda
——— v. fpeciofa.	*A new variety* of ditto
——— oblongata.	Long leaved Monarda
——— didyma.	Scarlet ditto, or Ofwego Tea
——— rugofa.	White ditto
——— clinopodea.	Bafil leaved ditto
Salvia Cretica.	Cretan fage
——— lyrata.	Lyre leaved ditto
——— Hablizliana.	Hablizlian ditto
——— officinalis.	Garden fage
——— v. fol. arg.	Silver blotched ditto
——— v. fol. aur.	Gold ditto
——— viridis.	Green toped fage
——— Horminum v. *coma*	Purple toped ditto
——— *violacea.*	
——— v. coma rub.	Red ditto
——— virgata.	Long branched ditto
——— fylveftris.	Spotted ftalked Bohemian ditto
	Salvia

Salvia nemorosa.	Spear leaved ditto
—— viscosa.	Clammy ditto
—— pratensis.	Meadow ditto
—— bicolor.	Two coloured ditto
—— Verbenaca.	Vervain sage, or clary
—— clandestina.	Cut-leaved ditto
—— Austriaca.	Austrian ditto
—— Nilotica.	Nilotic ditto
—— Hispanica.	Spanish ditto
—— verticillata.	Whorl flowered ditto
—— napifolia.	Rape-leaved ditto
—— glutinosa.	Yellow ditto
—— Sclarea.	Common clary
—— ceratophylla.	Horn-leaved sage
—— Æthiopis.	Woolly ditto
—— Forskaelei.	Forskael ditto
Collinsonia Canadensis.	Nettle leaved Collinsonia
Anciftrum sanguiforba.	Burnet leaved Anciftrum
—— lucidum.	Shining Anciftrum
—— latebrofum.	Hairy ditto

DIANDRIA DIGYNIA.

Anthoxanthum odoratum.	Sweet scented spring grass

CLASSIS III.

TRIANDRIA MONOGYNIA.

Valeriana rubra.	Red valerian
—— v. fl. alb.	White flowered ditto
—— calcitrapa.	Cut-leaved ditto
—— dioica.	Marsh ditto
—— officinalis.	Officinal valerian

Valeriana

Valeriana Phu.	Garden ditto
———— tripteris.	Three leaved ditto
———— montana.	Mountain ditto
———— tuberofa.	Tuberous rooted ditto
———— pyrenaica.	Pyrenean ditto
———— fupina.	Dwarf ditto
———— olitoria.	Corn ditto
Valeriana dentata.	Dentated valerian
———— veficaria.	Bladder cuped ditto
———— coronata.	Coronate ditto
Ortegia Hifpanica.	Spanifh Ortegia
Loeflingia Hifpanica.	Spanifh Lœflingia
Polycnemum arvenfe.	Trailing Polycnemum
Crocus fativus.	Saffron
———— nudiflorus.	Naked flowered ditto
———— vernus.	Blue flowered fpring Crocus
———————— v.	Yellow ditto
———————— v.	Afh coloured ditto
———————— v.	Cloth of gold ditto
Ixia Bulbocodium.	Crocus leaved Ixia
Gladiolus communis.	Common red corn flag
—————— v fl. alb.	White ditto
—————— v. fl. incar.	Flefh coloured ditto
—————— Byzantinus.	
Iris pumila.	Dwarf Iris
— lutefcens.	Yellowifh ditto
— criftata.	Crefted ditto
— Sufiana.	Chalcedonian ditto
— Florentina.	Florentine ditto
— biflora.	Two flowered ditto
— aphylla.	Naked ftalked ditto
— variegata.	Variegated ditto
— fqualens.	Brown flowered ditto
— fambucina.	Elder fcented ditto
— lutida.	Dingy ditto
— Germanica.	German ditto

H   Iris

Iris pallida.	Pale flowered ditto
— Swertia.	
— Xiphium.	Bulbous rooted ditto
— Xiphioides.	Englifh bulbous ditto
— Pfued-Acorus.	Yellow Iris
— fœtidiffima.	Stinking ditto
———— *v. fol. varieg.*	Variegated ditto
— Virginica.	Virginian ditto
— purpurafcens.	Purpurefcent ditto
— verficolor.	Various coloured ditto
— ocroleuca.	Pale yellow ditto
— Sifyrinchium.	Crocus rooted ditto
— Perfica.	Perfian ditto
— graminea.	Grafs leaved ditto
— fpuria.	Spurious ditto
— flexuofa.	Zigzag ditto
— Sibirica.	Siberian ditto
— tuberofa.	Snake's head ditto
Moræa Chinenfis.	Chinefe Moræa
Commelina erecta.	Upright Virginian Commelina
Schoenus marifcus.	Prickly bog rufh
———— nigricans.	Black ditto
———— rufus.	Tawny rufh grafs
———— fufcus.	Brown ditto
———— albus.	White flowered ditto
Cyperus longus.	Englifh galingale
———— *nova fpecies.*	
Scirpus paluftris.	Marfh club rufh
———— caricis.	Compreffed bog rufh
———— cæfpitofus.	Dwarf club rufh
———— atropurpureus.	
———— acicularis.	Leaft upright ditto
———— fluitans.	Floating ditto
———— lacuftris.	Bull-rufh
———— fetaceus.	Leaft club-rufh

Scirpus

Scirpus triqueter.	Pointed ditto
———— maritimus.	Sea ditto
———— fylvaticus.	Wood ditto
Eriophorum vaginatum.	Mountain cotton grafs
———————— polyftachion.	Broad-leaved ditto
———————— anguftifolium.	Common ditto
———————— alpinum.	Alpine ditto
Nardus ftri&ta.	Mat grafs
Lygeum fpartum.	Rufh leaved Lygeum

TRIANDRIA DIGYNIA

Cornucopiæ cucullatæ.	Hooded Cornucopiæ.
Leerfia oryzoides.	
Phalaris Canarienfis.	Manured Canary grafs
———— aquatica.	Water ditto
———— phleoides.	Naked cuped ditto
———— nodofa.	Knotted ditto
———— arenaria.	Sea ditto
———— afpera.	Rough ditto
———— paradoxa.	Briftly fpiked diitto
Panicum verticillatum.	Rough panic grafs
———— glaucum.	Glaucous panic grafs
———— viridi..	Green ditto
———— Italicum.	Italian ditto
———— ciliatum.	Ciliated ditto
———— crus-galli.	Thick fpiked cocks foot ditto
———— fanguinale.	Slender fpiked cocks foot ditto
———— da&tylon.	Creeping ditto
———— filiforme.	Clofe fpiked ditto
———— milliaceum.	Milet panic grafs
———— capillare.	Hair panicled ditto
Phleum pratenfe.	Cat's tail grafs
———— alpinum.	Mountain ditto
———— nodofum.	Bulbous ditto

Alopecurus bulbofus.	Bulbous foxtail grafs
———— pratenfis.	Meadow ditto
———— agreftis.	Field ditto.
———— geniculatus.	Flote ditto
Milium lendigeron.	Yellow fpiked millet grafs
——— effufum.	Common ditto
——— paradoxum.	Black feeded ditto
Agroftis fpica-venti.	Silky bent grafs.
——— panicea.	Bearded foxtail ditto
v.	
——— fetacea.	Briftly ditto
——— miliacea.	Millet bent grafs
——— rubra.	Red ditto
——— canina.	Brown bent ditto
——— littoralis	
——— ftolonifera.	Creeping ditto
——— hifpida.	Fine ditto
——— fylvatica.	Wood ditto
——— alba.	White ditto
——— paluftris.	Marfh ditto
——— maritima.	Sea ditto
——— pumila.	Dwarf ditto
——— tenuifolia.	Slender leaved ditto
——— minima.	Leaft, or fmall ditto
——— Mexicana.	Mexican ditto
——— verticilata.	Whorled ditto
Aira aquatica.	Water hair grafs
——— cæfpitofa.	Turfy ditto
——— flexuofa.	Heath ditto
——— canefcens.	Grey hair grafs
——— præcox.	Early ditto
——— caryophyllea.	Silvery leaved ditto
Melica ciliata.	Ciliated melic grafs
——— fericia.	Silky ditto
——— nutans.	Mountain ditto
——— uniflora.	Single flowered wood ditto

Melica

Melica cærulea.	Purple ditto
——— altiſſima.	Tall ditto
Poa aquatica,	Water meadow graſs
—— alpina.	Alpine ditto
—— trivialis.	Common ditto
—— anguſtifolia.	Narrow leaved ditto
—— pratenſis.	Great ditto
—— paluſtris.	Marſh ditto
—— annua.	Annual ditto
—— maritima.	Sea ditto
—— retroflexa.	Retroflexed ditto
—— diſtans.	Looſe flowered ditto
—— rigida.	Hard ditto
—— Eragroſtis.	Spreading ditto
—— rupeſtris.	Rock ditto
—— cæſia.	
—— flexuoſa.	
—— compreſſa.	Creeping ditto
—— Ceniſia.	Soft ditto
—— Amboinenſis.	
—— nemoralis.	Wood ditto
—— criſtata.	Creſted dito
Briza minor.	Small quaking graſs
—— virens.	Spaniſh ditto
—— media.	Middle ditto
—— maxima.	Great ditto
Uniola ſpicata.	Spiked Uniola
——— paniculata.	Panicled ditto
Dactylis cynoſuroides.	American cocks foot graſs.
—— patens.	Spreading ditto
—— glomerata.	Rough ditto
Cynoſurus criſtatus.	Creſted dogs tail graſs
——— echinatus.	Rough ditto
——— eruræformis.	Linear ſpiked ditto
——— durus.	Rigid ditto
——— cæruleus.	Blue ditto

Cynoſurus

Cynosurus aureus.	Golden spiked dog's tail grafs
Feſtuca bromoides.	Barren fefcue grafs
———— ovina.	Sheeps ditto
———— *v. vivipara.*	Viviparous ditto
———— rubra.	Red ditto
———— pumila.	Dwarf ditto
———— amethyſtina.	Amethyſtine ditto
———— duriuſcula.	Hard ditto
———— dumetorum,	Wood ditto
———— glauca.	Glaucous ditto
———— Cambrica.	Welſh ditto
———— myurus.	Wall ditto
———— calycina.	Bearded leaved ditto
———— unigluma.	Sea ditto
———— ſpadicea.	Spadiceus ditto
———— decumbens.	Decumbent ditto
———— elatior.	Tall ditto
———— calamaria.	
———— pratenſis.	Meadow ditto
———— fluitans.	Flote ditto
———— loliacea.	Darnel ditto
Bromus ſecalinus.	Field brome grafs
———— multiflorus.	Many flowered ditto
———— mollis.	Soft ditto
———— ſquarroſus.	Corn ditto
———— purgans.	Canadian.
———— inermis.	Smoth ditto
———— aſper.	Wood ditto
———— ciliatus.	Ciliated brome grafs
———— ſterilis.	Barren ditto
———— arvenſis.	Corn ditto
———— tectorum.	Nodding panicled ditto
———— giganteus.	Tall ditto
———— alopecurus.	Fox-tail
———— rubens.	Spaniſh ditto
———— erectus.	Upright ditto

Bromus

Bromus Madritenſis.	Wall ditto
—— gracilis.	Slender ditto
—— pinnatus.	Spiked ditto
—— ſylvaticus.	Hairy ſtalked ditto
—— ſtipoides.	Feathery ditto
Stipa pennata.	Soft feather graſs
—— juncea.	Ruſh leaved ditto
Avena Sibirica.	Siberian oat graſs
—— elatior.	Tall ditto
—— *v. nodoſa.*	Bulbous rooted ditto
—— *v. muticis.*	Awnleſs ditto
—— brevis.	Smooth ditto
—— ſtrigoſa.	Scrannnel ditto
—— ſativa v. *nigra.*	Cultivated black oats
—— *v. albu.*	Cultivated white ditto
—— nuda.	Naked oat graſs
—— fatua.	Wild ditto
—— ſeſquitertia.	
—— pubeſcens.	Soft ditto
—— ſterilis.	Bearded ditto
—— flaveſcens.	Yellow ditto
—— pratenſis.	Meadow ditto
Lagurus Ovatus.	Oval ſpiked Lagurus
Arundo Donax.	Manured reed graſs
—— *v. verſicolor.*	Striped ditto
—— Phragmitis.	Common reed graſs
—— Epigegos.	Small reed graſs
—— Calamagroſtis.	Wood ditto
—— colorata.	Reed canary ditto
—— colorata v. *fol. arg.*	Silver ſtriped reed canary graſs
—— *v. fol. aur.*	Gold ditto
—— arenaria.	Sea reed ditto
Lolium perenne.	Perennial darnel graſs
—— *v. compreſs.*	Flat headed ditto
—— *v. ramos.*	Branched ditto
—— tenue.	Slender ditto

Lolium.

Lolium temulentum.	Annual ditto
———————— *v. muticis.*	Awnlefs ditto
——— arvenfe.	White ditto
Rothboellia incurvata.	Sea hard grafs
Elymus arenarius.	Sea lime grafs
——— geniculatus.	Jointed ditto
——— giganteus.	Gigantic ditto
——— Sibiricus.	Siberian ditto
——— Philadelphicus.	Philadelphian ditto
——— Canadenfis.	Canadian ditto
——— caninus.	Dog's ditto
——— Virginicus.	Virginian ditto
——— Europeus.	Wood ditto
——— hyftrix.	Rough ditto
Secale cereale.	Winter manured rye
Hordeum vulgare.	Spring barley
———————— *v. cæleft.*	
———————— *v. fem. nigris.*	Black feeded ditto
——— hexaftichon.	Winter ditto
——— diftichon.	Common ditto
———————— *v. nudum.*	Naked ditto
——— Zeocriton.	Battledore ditto
——— murinum.	Wall ditto
——— fecalinum.	Meadow ditto
——— maritimum.	Sea ditto
——— jubatum.	Long bearded ditto
Triticum æftivum.	Summer wheat
——— hybernum.	Winter, or lamas ditto
——— compofitum.	Branched ditto
——— turgidum.	Turgid, or cone ditto
———————— *v. arift. nigr.*	Black awned ditto
——— Polonicum.	Poland ditto
——— fpelta.	Spelt ditto
——— monococcum.	One grained ditto
——— junceum.	Sea ditto

Tritieum

Triticum repens.   Dog's ditto, or couch grafs
———— tenellum   Dwarf wheat grafs
———— loliaceum.   Spiked fea ditto

TRIANDRIA- TRIGYNIA.

Montia fontana.   Water chickweed
Holofteum umbellatum.   Umbelled Holofteum
Polycarpon tetraphyllum. Four leaved Polycarpon

CLASSIS IV.

TETRANDRIA MONOGYNIA.

Globularia vulgaris.   Common Globularia
———— cordifolia.   Wedge leaved ditto
———— nudicaulis.   Naked ftalked ditto
Dipfacus Fullonum.   Wild teafel
———— fylveftris.   Manured ditto
———— laciniatus.   Cut-leaved ditto
———— pilofus.   Small ditto
Scabiofa alpina.   Alpine fcabious.
———— leucantha.   Snowy ditto
———— fuccifa.   Devil's bit ditto
———— Tartarica.   Giant ditto
———— arvenfis.   Field ditto
———— fylvatica.   Broad leaved ditto
———— gramuntia.   Cut-leaved fcabious
———— columbaria.   Fine-leaved ditto
———— ftellata.   Starry ditto
———— atropurpurea.   Sweet ditto
———— Ucranica.   Ucrania ditto
———— ocroleuca   Pale white ditto
———— pappofa.   Downy-headed ditto

I   Scabiofa

Scabiofa ficula.	Sicilian Scabious
Knautia Orientalis.	Oriental Knautia
Sherardia arvenfis.	Field madder
Afperula odorata.	Sweet fcented woodroof
———— arvenfis.	Field ditto
———— Tartarica.	Broad-leaved ditto
———— ariftata.	Thick-leaved ditto
———— tinctoria.	Narrow-leaved ditto
———— Cynanchica.	Small ditto, or fquinancy wort
———— lævigata.	Shining ditto
Houftonia cærulea.	Blue flowered Houftonia
Galium rubioides.	Madder leaved ladies bed-ftraw
———— paluftre.	White ditto
———— montanum.	Mountain ditto
———— pufillum.	Leaft ditto
———— erectum.	Upright ditto
———— verum.	Yellow ditto
———— mollugo.	Great ditto
———— fylvaticum.	Wood ditto
———— glaucum.	Glaucous ditto
———— fpurium.	Corn ditto
———— uliginofum.	Marfh ditto
———— boreale.	Crofs leaved ditto
———— rotundifolium.	Round-leaved ditto
———— Aparine.	Common ditto, or Clivers.
———— Parifienfe.	Small ditto
Crucianella anguftifolia.	Narrow-leaved Crucianella
———— patula.	Spreading ditto
Rubia tinctorum.	Dyer's madder
———— peregrina.	Wild ditto
Mitchella repens.	Creeping Mitchella
Plantago major.	Great plantain
———————— v. latif. rofea.	Rofe ditto
———— craffa.	Curled ditto
———— Afiatica.	Afiatic ditto
———— maxima.	Broad leaved ditto

Plantago

Plantago media.	Hoary plantain
———— Virginica.	Virginian ditto
———— altissima.	Tall ditto
———— lanceolata.	Ribwort ditto
———— Lagopus.	Round-headed ditto
———— alpina.	Alpine ditto
———— rigida.	Rigid ditto
———— maritima.	Sea ditto
———— subulata.	Awl-leaved ditto
———— Serraria.	
———— Coronopus.	Buck's-horn ditto
———— aomplexicaulis.	Stem clasping ditto
———— Psyllium.	Clammy ditto
———— squarrosa.	Scaly ditto
———— Cynops.	Shrubby ditto
Centunculus minimus.	Small Centunculus
Sanguisorba officinalis.	Common burnet saxifrage
———— media.	Short spiked ditto
———— Canadensis.	Canadian ditto
Epimedium alpinum.	Barren wort
Cornus Suecica.	Herbaceous dogwood
———— Canadensis.	Canadian ditto
Alchemilla vulgaris.	Common ladies mantle
———— v. hybrida.	Pubescent ditto
———— alpina.	Alpine ditto
———— pentaphylla.	Five-leaved ditto
———— aphanes.	Parsley piert

TETRANDRIA TETRAGYNIA.

Bufonia tenuifolia.	Slender leaved Bufonia.
Hypecoum procumbens.	Procumbent Hypecoum
Potamogeton natans.	Broad leaved pondweed
———— perfoliatum.	Perfoliate ditto
———— densum.	Forked ditto
———— lucens.	Shining ditto

Potamogeton crifpum.	Curled Pondweed
————— comprefſum.	Flat ſtalked ditto
————— gramineum.	Graſſy ditto
————— marinum.	Sea ditto
Ruppia maritima.	Sea taſſel graſs
Sagina procumbens.	Procumbent pearlwort
——— apetala.	Annual ditto
——— erecta.	Upright ditto
Tillæa muſcoſa.	Procumbent Tillæa

CLASSIS V.

PENTANDRIA MONOGYNIA.

Heliotropium Europeum.	European turnſole
Myoſotis ſcorpiodes.	Hairy mouſe-ear ſcorpion graſs
————— *v. paluſtris.*	Marſh ditto
————— Lappula.	Prickly-ſeeded ditto
————— nana.	Dwarf ditto
Lithoſpermum officinale.	Officinal gromwell
——— arvenſe.	Baſtard alkanet
——— Orientale.	Oriental alkanet
——— purpureo-cæruleum.	Creeping ditto
Anchuſa officinalis.	Officinal bugloſs
——— Italica.	Italian ditto
——— anguſtifolia v. *alb.*	White flowered narrow leaved ditto
————— *v. fl. rubente.*	Purple ditto
——— ſempervirens.	Evergreen alkanet
Cynogloſſum officinale.	Officinal hounds-tongue
————— *v. ſylvat.*	Wood ditto
————— pictum.	Madeira ditto
——— Virginicum.	Virginian ditto
——— cheirifolium.	Silvery leaved ditto
——— Apenninum.	Apennine ditto

Cynogloſſum

Cynoglossum angustifolium.	Narrow-leaved hounds-tongue
———— linifolium.	Flax-leaved ditto
—— —— omphalodes.	Comfrey-leaved ditto
Pulmonaria angustifolia.	Narrow-leaved lung-wort
———————— *v. fol. macul.*	Spotted narrow-leaved ditto
———— officinalis v. *fol. rub.*	Common ditto
——————— *v. fol. alb.*	White flowered ditto
——— —— *v. fol non macul.*	Unspotted ditto
———— Virginica.	Virginian ditto
———— maritima.	Sea ditto
Symphytum officinale.	Common comfrey
——— ———— *v. patens.*	Spreading ditto
———— tuberosum.	Tuberous ditto
Cerinthe major.	Great purple honeywort
——— ———— *v. fl. flav.*	Great yellow ditto
——— minor.	Small ditto
——— aspera.	Rough ditto
Borago officinalis.	Common borage
——— Orientalis.	Oriental ditto
Asperuga procumbens.	Great goose-grass
Lycopsis pulla.	Dark flowered wild buglofs.
——— arvensis.	Small wild ditto
Echium plantagineum.	Plantain leaved vipers buglofs
——— Italicum.	Wall ditto
——— vulgare.	Common ditto
——— violaceum.	Violet-flowered ditto
——— Creticum.	Cretan ditto
——— Orientale.	Oriental ditto
Nolana prostrata.	Trailing Nolana.
Aretia vitalina.	Grafs-leaved Aretia
Androsace maxima.	Oval-leaved Androsace
———— elongata.	Cluster-flowered ditto
———— Septentrionalis.	Tooth-leaved ditto
———— villosa.	Hairy leaved ditto
———— chamæjasme.	
———— lactea.	Grafs-leaved ditto

Primula

Primula veris.	Common primrose
———— *v. alb.*	White flowered ditto
———— *v. carnea.*	Flesh coloured ditto
———— *v. sulphur.*	Sulphur ditto
———— *v. velvetara.*	Velvet ditto
——— officinalis.	Common cowslip
———— *v. coccinea.*	Scarlet ditto
———— *v. plen.*	Double flowered ditto
——— elatior.	Oxslips
———— *v.*	Double polyanthus
——— farinosa.	Bird's-eye primrose
——— cortusoides.	Cortusa leaved ditto
——— villosa.	Villous ditto
——— nivalis.	Snowy ditto
——— marginata.	Edged ditto
——— Helvetica.	Swiss ditto
——— Auricula *v. fl. lutea.*	Yellow auricula
———— *v. fl. purp.*	Purple ditto
———— *v. fl. varieg.*	Variegated ditto
——— integrifolia.	Entire leaved ditto
——— Norwegica.	Norway ditto
——— Sibirica.	Siberian ditto
Soldanella Alpina.	Alpine Soldanella
Dodecatheon Meadia.	Virginian cowslip
Cyclamen Coum.	Round leaved Cyclamen
——— Europeum.	Common European ditto
——— hederæfolium.	Ivy-leaved ditto
Menyanthes nymphoides.	Fringed buck-bean, or water lilly
——— trifoliata.	Common buck-bean, or marsh trefoil
Hottonia palustris.	Water Violet
Hydrophyllum Virginicum.	Virginian water leaf
——— Canadense.	Canadian ditto
Lysimachia vulgaris.	Common loose strife
——— ephemerum.	Willow-leaved ditto

Lysimachia

Lyfimachia ftricta.	Upright loofe ftrife
———— thyrfifolia.	Tufted ditto
———— quadrifolia.	Four-leaved ditto
———— Linum-ftellatum.	Small ditto
———— nemorum.	Wood ditto
———— Nummularia.	Creeping ditto
Anagalis arvenfis v. *fl. cærul.*	Blue pimpernel
———— *v. phæniceo.*	Red ditto
———— latifolia.	Broad-leaved ditto
———— tenella.	Burple money-wort
Spigella Marilandica.	Perennial worm-grafs
Plumbago Europea.	European lead-wort
Phlox paniculata.	Panicled lychnidea
—— undulata.	Wave-leaved ditto
—— fuaveolens.	White flowered ditto
—— maculata.	Spotted ftalked ditto
—— pilofa.	Hairy flalked ditto
—— Carolina.	Carolina ditto
—— glaberrima.	Smooth ditto
—— divaricata.	Early flowering ditto
—— ovata.	Oval leaved ditto
—— fubulata.	Awl-leaved ditto
—— fetacea.	Briftle-leaved ditto
Convolvulus arvenfis.	Small bindweed
———— fepium.	Great ditto
———— Scammonia.	Scammony
———— hederaceus.	Ivy-leaved bindweed
———— purpureus.	Great purple ditto
———— *v. fl. alb.*	Great white ditto
———— ficulus.	Small flowered ditto
———— lineatus.	Dwarf ditto
———— tricolor.	Trailing ditto
———— *v. fl. alb.*	White flowered ditto
———— Soldanella.	Sea ditto
Ipomæa coccinea.	Scarlet flowered Ipomea

Polemonium

Polemonium cæruleum.	Blue flowered Greek valerian
———————— *v. fl. alb.*	White ditto
——— reptans.	Creeping ditto
Campanula rotundifolia.	Round leaved bell flower
——— pumila.	Dwarf ditto
——— cæfpitofa.	Heathy fmooth ditto
——— carpatica.	Heart-leaved ditto
——— Lobelioides.	Small flowered ditto
——— patula.	Spreading ditto
——— rapunculus.	Efculent ditto, or rampions
——— perficifolia.	Single blue peached-leaved bell
——— *v. fl. cærul. plen.*	Double flowered ditto [flower
——— *v. fl. alb. fimp.*	Single white ditto
——— *v. fl. plen.*	Double white ditto
——— pyramidalis.	Pyramidal ditto
——— nitida.	Smooth leaved ditto
——— lilifolia.	Lilly ditto
——— rhomboidea.	
——— latifolia.	Broad leaved ditto
——— rapunculoides.	Nettle-leaved ditto
——— trachelium	Great blue flowered ditto
——— *v. fl. plen.*	Great double blue flowered do.
——— *v. fl. alb.*	Great white ditto
——— *v. fl. pleno.*	Great double white ditto
——— glomerata.	Cluftered ditto
——— thyrfoidea.	Thyrfe ditto
——— peregrina.	Round leaved ditto
——— medium.	Blue Canterbury bell-flower
——— *v. fl. alb.*	White ditto
——— barbata.	Bearded ditto
——— fpeculum.	Venus's looking-glafs
——— hybrida.	Corn bell flower
——— pentagonia	
——— perfoliata.	Perfoliate ditto
——— hederacea.	Ivy-leaved ditto

Campanula

Campanula verticillata.	Whorled bell flower
———— erinus.	Forked ditto
Phyteumea orbicularis.	Round headed horned rampion
———— fpicata.	Spiked ditto
———— lanceolata.	Spear leaved ditto
Trachelium cæruleum.	Blue throat wort
Samolus valerandi.	Round leaved water-pimpernel
Triofteum perfoliatum.	Fever-root
Mirabilis Jalapa.	Common marvel of Peru
———— *v. fl. flavo.*	Yellow ditto
———— *v. fl. varieg.*	Yellow ftriped ditto
———— *v. fl. alb.*	White ditto
———— longiflora.	Sweet fcented ditto
Verbafcum Thapfus.	Great broad leaved mullein
———— *v. fl. alb.*	Great white ditto
———— phlomoides.	Woolly ditto
———— lychnitis.	Small yellow flowered ditto
———— *v. fl. alb.*	Small white ditto
———— *v. pulverulentum.*	
———— ferrugineum.	Rufty ditto
———— nigrum.	Black ditto
———— Phœniceum.	Purple ditto
———— Blattaria.	Yellow flowered moth ditto
———— *v. fl. alb.*	White flowered moth ditto
———— virgatum.	Rod-like ditto
———— Myconi.	Borage leaved ditto
Datura Stramonium.	Common thorn apple
———— Tatula.	Blue ditto
Hyofcyamus niger.	Black henbane
———— albus.	White ditto
———— phyfaloides.	Purple flowered ditto
———— Scopolia.	Night-fhade leaved ditto
Nicotiana Tabacum.	Narrow-leaved Virginian tobacco
———— *v. latif.*	Broad-leaved ditto
———— ruftica.	Common ditto

K

Nicotiana

Nicotiana paniculata.	Panicled tobacco
———— glutinofa.	Clammy-leaved ditto
Atropa Mandragora.	Mandrake
———— Belladona.	Deadly nightfhade
———— phyfaloides.	Blue flowered ditto
Phyfalis Alkekengi.	Common winter cherry
———— angulata.	Toothed leaved ditto
———— chenopodifolia.	Crack leaved ditto
Solanum tuberofum.	Common potato
———— Lycoperficum.	Love apple
———— nigrum.	Common nightfhade
———— *v. villofum.*	Yellow berried ditto
Capficum annuum.	Long podded Capficum
Chironia Centaurium.	Leffer centory
Claytonia Sibirica.	Siberian Claytonia.
———— perfoliata.	Perfoliate ditto
Illecebrum capitatum.	Headed knot grafs
———— verticillatum.	Whorled ditto
Glaux maritima.	Sea milk-wort
Tabernæmontana Amfonia.	Alternate leaved Tabernæmontana.
———— anguftifolia.	Narrow leaved ditto

PENTANDRIA DIGYNIA.

Cynanchum acutum.	Acute leaved Cynanchum
Apocynum androfæmifolium.	Tutfan-leaved dog's-bane
———— cannabinum.	Hemp ditto
———— hypericifolium.	St. John's-wort-leaved ditto
Afclepias Syriaca.	Syrian fwallow wort
———— purpurafcens.	Purple Virginian ditto
———— amæna.	Oval-leaved ditto
———— incarnata.	Flefh coloured ditto
———— Vincetoxicum.	White officinal ditto
———— nigra.	Black ditto
———— tuberofa.	Tuberous-rooted ditto

Herniaria

Herniaria glabra.	Smooth rupture-wort
———— hirſuta.	Hairy ditto
———— lenticulata.	Sea ditto
Chenopodium Bonus-Henricus.	Engliſh mercury
———— urbicum.	Upright gooſe-foot
———— rubrum.	Red ditto
———— murale.	Wall ditto
———— ſerotinum.	Fig-leaved ditto
———— album.	Common ditto
———— viride.	Green ditto
———— hybridum.	Baſtard ditto
———— botrys.	Cut-leaved ditto
———— ambroſioides.	Mexican ditto
———— cordatum.	Heart-leaved ditto
———— glaucum.	Oak-leaved ditto
———— Vulvaria.	Stinking ditto
———— polyſpermum.	Round leaved ditto, or allſeed
———— Scoparia.	Linear-leaved ditto, or ſummer cypreſs
———— maritimum.	Sea ditto
———— ariſtatum.	Bearded ditto
Beta vulgaris.	Red beet
———— *v. virens*	Green ditto
——— Cicla.	White ditto
——— maritima.	Sea ditto
Salſola Kali.	Prickly ſaltwort
——— roſacea.	Roſe ditto
——— altiſſima.	Graſs-leaved ditto
Heuchera Americana.	American ſanicle
Swertia perennis.	Marſh Swertia
Gentiana lutea.	Yellow gentian.
——— purpurea.	Purple ditto
——— punctata.	Spotted flowered gentian
——— aſclepiadea.	Swallow wort leaved ditto
——— cruciata.	Croſs-wort ditto

Gentiana

Gentiana pneumonanthe	Marsh ditto, or calathian violet
———— saponaria.	Soap-wort leaved ditto
———— acaulis.	Dwarf ditto, or gentianella.
———— verna.	Spring ditto
———— amarella.	Autumnal ditto
———— campestris.	Field ditto
Eryngium aquaticum.	Marsh eryngo
———— planum.	Flat-leaved ditto
———— maritimum.	Sea ditto
———— campestre.	Common ditto
———— amethystinum.	Amethystian ditto
———— alpinum.	Alpine ditto
———— Bourgati.	Cut-leaved ditto
Hydrocotyle vulgaris.	Common marsh penny-wort
———— Americana.	American ditto
Sanicula Europea.	Common sanicle.
———— Marilandica	Maryland ditto
Astrantia major.	Great black master wort
———— minor.	Small ditto
Bupleurum rotundifolium.	Round-leaved hares ear
———— petræum.	Rock ditto
———— longifolium.	Long leaved ditto
———— ranunculoides.	Spear-wort leaved ditto
———— tenuissimum.	Slender ditto
———— junceum.	Linear leaved ditto
Hasselquistia cordata.	Heart leaved Hasselquistia
Tordylium Syriacum.	Syrian hartwort
———— officinale.	Officinal ditto
———— maximum.	Great ditto
Caucalis daucoides.	Carrot-leaved Caucalis
———— latifolia.	Broad-leaved ditto
———— leptophylla.	Fine leaved ditto
———— arvensis.	Corn ditto.

Caucalis

Caucalis Anthrifcus.	Hedge ditto
——— nodofa.	Knotted ditto
Daucus Carota.	Wild carrot
———— Vifnaga.	Spanifh ditto
———— muricatus.	Prickly-feeded ditto
Ammi majus.	Bifhops weed
Bunium Bulbocaftanum.	Earth nut
Conium maculatum.	Common hemlock
Selinum paluftre.	Milk parfley
Athamanta libanotis.	Pyrenean mountain fpignel.
Peucedanum officinale.	Common fulphur-wort
———— Silaus.	Meadow ditto
Crithmum maritimum.	Sea famphire
Ferula communis.	Common gigantic fennel
—— glauca.	Glaucous ditto
—— Tingitana.	Tangier ditto
—— nodiflora.	Knotted ditto
—— Perfica.	Perfian ditto
Laferpitium latifolium.	Broad leaved lafferwort
———— trilobum.	Columbine leaved ditto
———— anguftifolium.	Narrow-leaved ditto
———— Siler.	Mountain ditto
———— fimplex.	Small ditto
Heracleum Sphondylium.	Common cow parfnip
———— longifolium.	Long leaved ditto
———— Sibiricum.	Siberian ditto
———— alpinum.	Mountain ditto
———— flavefcens.	Yellowifh ditto
———— anguftifolium.	Narrow leaved ditto
Ligufticum Levifticum.	Common lovage
———— Scoticum.	Scotch ditto
———— Peloponnenfe.	Hemlock-leaved ditto
———— Cornubienfe.	Cornifh ditto
———— candicans.	Pale ditto
Angelica archangelica.	Garden angelica
———— fylveftris.	Wild ditto

Angelica

Angelica verticillaris.	Whorl flowered ditto
—— —— atropurpurea.	Dark purple ditto
Sium latifolium.	Great water parfnip
—— anguftifolium.	Narrow-leaved ditto
—— nodiflorum.	} Creeping ditto
—— repens.	
—— Sifarum.	Skirret
—— Falcaria.	Decurrent ditto
Sifon Amomum.	Field hone-wort
—— fegetum.	Corn ditto
—— inundatum.	Water ditto
—— verticillatum.	Whorled leaved ditto
Cuminum Cyminum.	Cumin
Œnanthe fiftulofa.	Common water drop-wort
—————— crocata.	Hemlock ditto
—————— peucedanifolia.	Sulphur-wort leaved ditto
—————— pimpinelloides.	Parfley leaved ditto
Phellandrium aquaticum.	Water hemlock
—————Mutellina.	Alpine Phellandrium
Cicuta virofa.	Long leaved water hemlock
Æthufa Cynapium.	Common fools parfley
—— Meum.	Common fpignel
Coriandrum fativum.	Common coriander
Scandix odorata.	Sweet fcented cicely
—— Pecten.	Corn ditto, or fhepherds needle
—— cerefolia.	Garden ditto, or chervil
—— Anthrifcus.	Rough ditto
—— nodofa.	Knotted ditto
Chærophyllum fylveftre.	Wild chervil, or cow weed
—————— bulbofum.	Bulbous rooted Chærophyllum
—————— temulum.	Rough ditto
—————— hirfutum.	Hairy ditto
—————— aromaticum.	Aromatic ditto
—————— aureum.	Golden ditto
Imperatoria Oftruthium.	Common mafterwort

Sefeli

Seseli montanum.	Long leaved meadow saxifrage
—— aristatum.	Bearded leaved ditto
—— Hippomarathrum.	Various-leaved ditto
Thapsia villosa.	Deadly carrot
Pastinaca sativa.	Common parsnip
———— v. latif.	Broad leaved ditto
——— Opoponax.	Rough ditto
Smyrnium perfoliatum.	Perfoliate alexanders
———— Olusatrum.	Common ditto
———— aureum.	Golden ditto
Anethum graveolens.	Common dill
——— Fœniculum.	Common fennel
Carum Carui.	Common caraway
Pimpinella saxifraga.	Small burnet saxifrage
———— magna.	Great ditto
——— dissecta.	Cut leaved ditto
—— —— dioica.	Least pimpinell
Apium Petroselinum.	Common parsley
———— v. crisp.	Curled ditto
——— graveolens.	Common smallage
——— v. dulce.	Cellery
Aegopodium Podagraria.	Gout-weed

PENTANDRIA TRIGYNIA.

Sambucus Ebulus.	Dwarf elder
Telephium Imperati.	True orpine
Corrigiola litoralis.	Bastard knot grass
Alsine media.	Chickweed

PENTANDRIA TETRAGYNIA.

Parnassia palustris.	Grass of Parnassus

PENTANDRIA PENTAGYNIA.

Aralia racemosa.	Berry-bearing Aralia
—— nudicaula.	Naked stalked ditto

Statice

Statice Armeria.	Great common thrift
———— v. minor.	Small ditto
—— Cephalotes.	Large simple stalked ditto
—— Limonium.	Common sea ditto
—— oleæfolia.	Olive-leaved ditto
—— cordata.	Blunt leaved ditto
—— reticulata.	Matted ditto
—— speciosa.	Plantain leaved ditto
—— Tartarica.	Tartarian ditto
Linum usitatissimum.	Common flax
———— Tartaricum.	Tartarian ditto
——— perenne.	Perrenial ditto
——— tenuifolium.	Narrow leaved ditto
——— alpinum.	Alpine ditto
——— Austriacum.	Austrian ditto
——— flavum.	Yellow ditto
——— catharticum.	Purging ditto
——— Radiola.	Least ditto, or all-seed
Drosera rotundifolia.	Round leaved sun-dew
——— longifolia.	Long leaved ditto
Crassula rubens.	Annual Crassula
Sibaldia procumbens.	Trailing Sibaldia

PENTANDRIA POLYGYNIA.

Myosurus minimus.	Mouse-tail

CLASSIS VI,

HEXANDRIA MONOGYNIA.

Tradescantia Virginica.	Common blue Virginian spider-wort
———— ——— v. purp.	Purple ditto
——— ——— v. alb.	White ditto
Galanthus nivalis.	Common single snow-drop
——— ——— v. fl. pleno.	Double ditto

Leucojum

Leucojum vernum.	Great fpring ditto
———— æftivum.	Summer ditto
Narciffus anguftifolius.	Narrow leaved Narciffus
———— biflorus.	Two flowered ditto
———— majalis.	
———— *v. fl. pleno.*	
———— incomparabilis.	Peerlefs daffodil
———— *v. fl. pleno.*	Double flowered ditto
———— Pfuedo-Narciffus.	Single daffodil
———— *v. fl. pleno.*	Double ditto
———— major.	Single great Narciffus
———— *v. fl. pleno.*	Double ditto
———— minor.	Small ditto
———— triandrus.	Rufh-leaved ditto
———— Orientalis.	Oriental ditto
———— Tazetta.	Polyanthus ditto
———— odorus.	Sweet fcented ditto
———— Bulbocodium.	Hoop peticoat ditto
———— Jonquilla.	Single Jonquill
———— *v. fl. pleno.*	Double ditto
Pancratium maritimum.	Sea Pancratium, or daffodil
Amaryllis lutea.	Yellow Amaryllis
———— Atamafco.	Atamafco lily
———— Belladonna.	Belladonna ditto
Bulbocodium vernum.	Spring flowering Bulbocodium
Allium Ampeloprafum.	Great round-headed garlick
———— Porrum.	Common leek
———— victoriale.	Long rooted garlick
———— fubhirfutum.	Hairy ditto
———— obliquum.	Oblique-leaved ditto
———— rofeum.	Rofe ditto
———— Tartaricum.	Tartarian ditto
———— fativum.	Cultivated ditto
———— Scorodoprafum.	Rocambole ditto
———— arenarium.	Sand ditto
———— carinatum.	Mountain ditto

L   Allium

Allium Sphœrocephalon.	Small round-headed garlick
—— defcendens.	Purple headed ditto
—— pallens.	Pale flowered ditto
—— paniculata.	Panicled ditto
—— vineale.	Crow ditto
—— oleraceum.	Purple ftriped ditto
—— nutans.	Flat ftalked ditto
—— Afcolonicum.	Afcalonian ditto, or fhallot
—— fenefcens.	Narciffus leaved ditto
—— angulofum.	Angular ftalked ditto
—— nigrum.	Broad leaved ditto
—— urfinum.	Ramfon ditto
—— triquetrum.	Triangular ditto
—— Cepa.	Common onion
—— Moly.	Yellow garlick
—— fiftulofum.	Welfh onion
—— Schœnoprafum.	Common chives
—— Sibirieum.	Siberian garlick
Lilium candidum.	Common white lily
—— v. fol. varieg.	Variegated ditto
—— bulbiferum.	Bulb-bearing orange ditto
—— v phæniceum.	Purple bulb-bearing ditto
—— v minus.	Small ditto
—— pomponium.	Pomponian lily
—— Chalcedonicum.	Scarlet martagon ditto
—— fuperbum.	Great yellow martagon ditto
—— Martagon.	Purple ditto
—— Canadenfe.	Canada ditto
Fritillaria imperialis.	Crown imperial
—— v. fl. varieg.	Variegated ditto
—— Perfica.	Perfian fritillary, or lily
—— Pyrenaica.	Pyrenian ditto
—— Meleagris.	Common ditto
—— v fl. alb.	White flowered ditto
Uvularia ample ifolia.	Heart-leaved Uvularia
—— perfoliata.	Perfoliate ditto

                                    Erythronium

Erythronium Denſ-canis.	Dog's-tooth violet
———— v. fl. alb.	White flowered ditto
Tulipa ſylveſtris.	Italian yellow tulip
——— Geſneriana.	Common ditto
——— Breyniana.	Cape ditto
Ornithogalum luteum.	Yellow ſtar of Bethlehem
——— —— Pyrenaicum.	Pyrenean ditto
——————— latifolium.	Broad leaved ditto
——————— umbellatum.	Umbelled ditto
——————— nutans.	Neapolitan ditto
Scilla Italica.	Italian ſquill
——— Peruviana.	Blue flowered Peruvian ditto
———— — v. fl. alb.	White ditto
——— campanulata.	Spaniſh ditto
——— amœna.	Nodding ditto
——— bifolia.	Two leaved ſpring ditto
———— —— v. fl alb.	White flowered ditto
——— autumnalis.	Autumnal ditto
——— verna.	Small ditto
——— præcox.	Early ditto
Aſphodelus luteus.	Yellow aſphodel
———— ramoſus.	Branchy ditto
Anthericum ramoſum.	Brancky Anthericum
———— Liliago.	Graſs leaved ditto
———— liliaſtrum.	Savoy ditto
———— annuum.	Annual ditto
———— oſſifragum.	Lancaſhire ditto
———— calyculatum.	Scotch ditto
Aſparagus officinalis.	Common Aſparagus
Dracæna borealis.	Oval leaved Dracæna
Convallaria majalis.	Sweet ſcented lily of the valley
————— v. fl. pleno.	Double flowered ditto
————— v. fl. rubro.	Red flowered ditto
——— Japonica.	Graſs leaved ditto
——— verticillata.	Whorl leaved ditto
——— Polygonatum	Common Solomon's ſeal

Convallaria multiflora.	Broad leaved lily of the valley
———— racemofa.	Cluftered flowered ditto
———— ftellata.	Star flowered ditto
———— bifolia.	Leaft ditto
Hyacinthus non fcriptus.	Common hyacinth
———— *v. fl. alb.*	White flowered ditto
———— *v. fl. rub.*	Red ditto
———— Orientalis.	Garden ditto
———— Romanus.	Roman grape ditto
———— Mufcari.	Mufk ditto
———— monftrofus.	Feathered ditto
———— comofus.	Purple grape ditto
———— botryoides.	Blue grape ditto
———— *v. fl. alb.*	White flowered ditto
———— racemofus.	Cluftered ditto
Aletris Uvaria.	Great orange flowered Aletris
Yucca gloriofa.	Superb Adams needle
——— filamentofa.	Thready ditto
Hemerocallis flava.	Yellow day lily
———— fulva.	Copper-coloured ditto
———— minor.	Small ditto
Acorus Calamus.	Sweet Acorus, or flag
——— gramineus.	Chinefe fweet grafs
Juncus acutus.	Sea rufh
——— conglomeratus.	Round headed ditto
——— effufus.	Common foft ditto
——— glaucus.	Hard ditto
——— filiformis.	Leaft foft fea rufh
——— trifidus.	Three flowered ditto
——— fquarrofus.	Mofs ditto
——— articulatus.	Jointed ditto
——— compreffus.	Flat ftalked ditto
——— bulbofus.	Bulbous ditto
——— uliginofus.	Boggy ditto
——— Jacquini.	Jacquin's ditto
——— biglumis.	Two flowered ditto

Juncus

Juncus triglumis.	Three flowered rush
———— pilosus.	Hairy ditto
———— sylvaticus.	Wood ditto
———— niveus.	White flowered ditto
———— campestris.	Field ditto
———— spicatus.	Spiked ditto
Frankenia lævis.	Smooth Frankenia, or sea heath
———— purverulenta.	Dusty ditto
Peplis Portula.	Water purslane

HEXANDRIA TRIGYNIA.

Rumex Patientia.	Patience dock
———— sanguineus.	Bloody ditto
———— crispus.	Curled ditto
———— dentatus.	Dentated ditto
———— maritimus.	Sea ditto
———— acutus.	Sharp pointed ditto
———— obtusifolius.	Blunt leaved ditto
———— pulcher.	Fiddle ditto
———— vesicarius.	Bladder ditto
———— scutatus.	French sorrel
———— digynus.	Mountain dock
———— alpinus.	Alpine ditto
———— Acetosa.	Common sorrel
———— Acetosella.	Sheeps ditto
———— aculeatus.	Prickly-seeded ditto
———— Hydrolopathum.	Great water dock
———— giganteus.	Giant dock
Triglochin palustre.	Marsh arrow-headed grass
———— maritimum.	Sea ditto
Melanthium lætum.	Spear leaved Melanthium
———— Virginicum.	Virginian ditto
Trillium cernuum.	Stalked flowered Trillium
———— sessile.	Sessile flowered ditto
Colchicum autumnale.	Common meadow saffron

Colchicum

Colchicum *v. fl. pleno.*	Double flowered meadow faffron
——— variegatum.	Variegated ditto
Helonias bullata.	Spear-leaved Helonias
——— afphodeloides.	Grafs leaved ditto

HEXANDRIA POLYGYNIA.

Alifma Plantago.	Greater water plantain
——— Damafonium.	Star-headed ditto
——— ranunculoides.	Leffer ditto

CLASSIS VII.

HEPTANDRIA MONOGYNIA.

Trientalis Europea.	Chickweed winter green

HEPTANDRIA TETRAGYNIA.

Saururus cernuus.	Lizard's-tail

CLASSIS VIII.

OCTANDRIA MONOGYNIA

Tropæolum minus.	Small Indian crefs
——— majus.	Great ditto
——— *v. fl. pleno.*	Double flowered ditto
Rhexia Virginica.	Hairy leaved Rhexia
Oenothera biennis.	Broad-leaved tree primrofe
——— grandiflora.	Great flowered Oenothera
——— parviflora.	Small flowered ditto
——— muricata.	Prickly ditto
——— longiflora.	Long flowered ditto
——— molliffima.	Soft ditto
——— fruticofa.	Shrubby ditto
——— finuata.	Scollop-leaved ditto

Oenothera

Oenothera odorata.   Sweet scented Oenothera
———— purpurea.   Purple ditto
———— rosea.   Rose flowered ditto
———— pumila.   Dwarf ditto
———— tetraptera.   Winged podded ditto
Gaura Biennis.   Biennial Gaura
Epilobium angustifolium.   Rose-bay willow-herb
———— ———— v. fl. alb.   White flowered ditto
———— angustissimum.   Linear leaved Epilobium
———— hirsutum.   Large flowered ditto
———— montanum.   Mountain ditto
———— tetragonum.   Square stalked ditto
———— roseum.   Rose ditto
———— palustre.   Marsh ditto
———— alpinum.   Alpine ditto
———— cordifolium.   Heart leaved ditto
Chlora perfoliata.   Yellow wort

OCTANDRIA DIGYNIA.

Moehringia muscosa.   Mossy Moehringia

OCTANDRIA TRIGYNIA.

Polygonum Bistorta.   Great bistort
———— viviparum.   Small ditto
———— Virginianum.   Virginian Polygonum
———— amphibium.   Amphibious ditto
———— ocreatum.   Spear leaved ditto
———— Hydropiper.   Water ditto
———— Persicaria.   Spotted ditto
———— Pensylvanicum.   Pensylvanian ditto
———— Orientale.   Red flowered Oriental persicaria
———— aviculare.   Common knot grass
———— ———— v. marit.   Sea ditto
———— divaricatum.   Divaricated Polygonum
———— undulatum.   Waved ditto

                  Polygonum

Polygonum Tartaricum.	Tartarian Polygonum
———— Fagopyrum.	Buck wheat
———— Convolvulus.	Black bindweed
———— fcandens.	Climbing Polygonum

OCTANDRIA TETRAGYNIA.

Paris quadrifolia.	Herb Paris
Adoxa mofchatellina.	Tuberous mofchatel

CLASSIS IX.

ENNEANDRIA MONOGYNIA.

Rheum Rhaponticum.	Rhapontic rhubarb
———— palmatum.	Officinal ditto
———— compactum.	Thick-leaved ditto
———— undulatum.	Waved-leaved ditto
———— Tartaricum.	Tartarian ditto
———— hybridum.	Baftard ditto

ENNEANDRIA HEXAGYNIA.

Butomus umbellatus.	Flowering rufh

CLASSIS X.

DECANDRIA MONOGYNIA.

Sophora alopecuroides.	Foxtail Sophora
———— Auftralis.	Blue ditto
———— tinctoria.	Dyer's ditto
———— flavefcens.	Siberian ditto
———— lupinoides.	Lupin-leaved ditto
Caffia Marilandica.	Maryland Caffia
Dictamnus albus.	White fraxinella
———— v. fl. rub.	Red ditto

Ruta

Ruta graveolens.	Common rue
Zogophyllum Fabago.	Common bean caper
Gaultheria procumbens.	Trailing Gaultheria
Epigæa repens.	Creeping Epigæa
Pyrola minor.	Leſſer winter green
——— rotundifolia.	Round-leaved ditto

DECANDRIA DIGYNIA.

Chryſoſplenium alternifolium.	Alternate-leaved golden ſaxifrage
——————— oppoſitifolium.	Oppoſite-leaved ditto
Saxifraga Cotyledon	Pyramidal ſaxifrage
——— ligulata.	Tongued ditto
——— roſularis.	Roſe ditto
——— Aizoon.	Margined ditto
——— Penſylvanica.	Penſylvanian ditto
——— ſtellaris.	Starry ditto
——— craſſifolia.	Thick-leaved ditto
——— cordifolia.	Heart-leaved ditto
——— nivalis.	Snowy ditto
——— ſarmentoſa.	China ditto
——— umbroſa.	London pride
——— cuneifolia.	Wedge-leaved ſaxifrage
——— Geum.	Kidney-leaved ditto
——— oppoſitifolia.	Purple-flowered ditto
——— aſpera.	Rough ditto
——— Hirculus.	Marſh ditto
——— aizoides.	Yellow mountain ditto
——— rotundifolia.	Round-leaved ditto
——— granulata.	Single white ditto
—— granulata *v. fl. pl.*	Double white ſaxifrage
——— geranioides.	Crane-bill leaved ditto
——— ajugifolia.	Bugle-leaved ditto
——— hypnoides.	Moſs ditto
—— viſcoſa.	Clammy ditto
—— tridactylites.	Rue-leaved ditto

Saxifraga petræa.	Rock faxifrage
———— cæfpitofa.	Matted ditto
———— Groenlandica.	Greenland ditto
Tiarella cordifolia.	Heart-leaved Tiarella
Mitella dipylla.	Two leaved Mitella
Scleranthus annuus.	Annual knawell
———— perennis.	Perrenial ditto
Gypfophilla repens.	Creeping Gypfophilla
———— proftrata.	Trailing ditto
———— vifcofa.	Clammy ditto
———— paniculata.	Panicled ditto
———— perfoliata.	Perfoliate ditto
———— muralis.	Wall ditto
———— Saxifraga.	Small ditto
Saponaria officinalis.	Common foap-wort
———— v. fl. pleno.	Double-flowered ditto
———— v. hybrid.	Hollow-leaved ditto
———— vaccaria.	Perfoliate ditto
———— porrigens.	Hairy ditto
———— ocymoides.	Bafil-leaved ditto
Dianthus barbatus.	Common fweet William
———— hybridus.	Hybrid ditto
———— Carthufianorum.	Carthufian pink
———— repens.	Creeping ditto
———— armeria.	Deptford ditto
———— proliferus.	Proliferous ditto
———— caryopylleus.	Clove ditto
———— v. fl. pleno.	Double ditto
———— deltoides.	Maiden ditto
———— glaucus.	Mountain ditto
———— cæfius.	Grey ditto
———— alpinus.	Mountain ditto
———— Chinenfis.	China ditto
———— plumarius.	Feathered ditto
———— fupetbus.	Superb ditto

Dianthus

Dianthus Virgineus.	Upright pink
—— pungens.	Pungent ditto

DECANDRIA TRIGYNIA.

Cucubalus Behen	Common bladder campion
———————— *v. marit.*	Sea ditto
———— bacciferus.	Berry-bearing ditto
———— Fabarius.	Thick-leaved ditto
———— viscosus.	Clammy ditto
———— Italicus.	Italian campion
———— Tartaricus.	Hyssop-leaved ditto
———— Catholicus.	Panicled ditto
———— Otites.	Spanish ditto
Silene Anglica.	English cathfly
—— Lusitanica.	Portugal ditto
—— quinquevulnera.	Mountain ditto
—— nocturna.	Spiked night flowering ditto
—— geniculiflorum.	Knotty flowered ditto
—— nutans.	Nottingham ditto
—— viridiflora.	Green flowered ditto
—— conidea.	Corn ditto
—— conica.	Conic ditto
—— bellidifolia.	Daisy leaved ditto
—— pendula.	Pendulous ditto
—— maritima.	Sea ditto
—— noctiflora.	Forked night flowering ditto
—— antirrhina.	Snap-dragon leaved ditto
—— inaperta.	Small flowered ditto
—— Portensis.	Oporto ditto
—— Cretica.	Cretan ditto
—— muscipula.	Fly trap ditto
—— Armeria.	Lobels red ditto
———————— *v. fl. alb.*	Lobels white ditto
—— orchidea.	Orchis flowered ditto
—— alpestris.	Austrian ditto

M 2   Silene

Silene rupeſtris.	Rock ditto
—— longiflora.	Long flowered ditto
—— bifidia.	Bifid ditto
—— ſaxifraga.	Saxifrage ditto
—— Vaileſia.	Woolly leaved catchfly
—— acaulis.	Dwarf ditto
Stellaria nemorum.	Broad leaved ſtitchwort
—— ſcapigera.	Naked ſtalked ditto
—— holoſtea.	Great ditto
—— graminea.	Graſs leaved ditto
—— ceraſtoides.	Small ditto
—— glauca.	Glaucous ditto
—— uliginoſa.	Fountain chickweed
Arenaria peploides.	Sea ſandwort
—— tetraquetra.	Square ditto
—— trinervia.	Plantain leaved ditto
—— Balearica.	Balearic ditto
—— ſerpyllifolia.	Thyme-leaved ditto
—— rubra.	Field purple ditto
—— v. marina.	Sea purple ditto
—— media.	Downy ditto
—— ſaxatilis.	Saxatile ditto
—— verna.	Spring ditto
—— tenuifolia.	Fine-leaved ditto
—— laricifolia.	Larch-leaved ditto
—— ſtriata.	Striated ditto
—— faſciculata.	Cluſter-flowering ditto
Cherleria ſedoides.	Stonecrop Cherleria
Garidella Nigellaſtrum.	Fennel-leaved Garidella

DECANDRIA PENTAGYNIA.

Cotyledon Umbilicus.	Common navelwort
—— lutea.	Yellow ditto
Sedum Telephium.	Great purple orpine
—— maximum.	Great white ditto

Sedum

Sedum Anacampſeros.	Evergreen orpine
—— Aizon.	Yellow ſtonecrop
—— hybridum.	Germander leaved ditto
—— populifolium.	Poplar leaved ditto
—— ſtellatum.	Starry ditto
—— cepæa.	Purſlane leaved ditto
—— dificiens.	American ditto
—— daſpyhyllum.	Round-leaved ditto
—— reflexum.	Reflexed ditto
—— virens.	Green ditto
—— glaucum.	Glaucous ditto
—— rupeſtre.	Rock ditto
—— album.	White ditto
—— acre.	Wall ditto
———— v.	
—— Anglicum.	Engliſh ditto
—— ſexangulare.	Six angled ditto
—— villoſum.	Hairy ditto
Penthorum ſedoides.	American Penthorum
Oxalis Acetoſella.	Common wood ſorrel
—— violacea.	Violet coloured ditto
—— corniculata.	Yellow ditto
—— ſtricta.	American ditto
—— Dillenii.	Dillenius's ditto
Agroſtemma Githago.	Cockle
———— Coronaria.	Red roſe campion
———— v. fl. alb.	White ditto
———— v. fl. pleno.	Double ditto
———— Flos-Jovis.	Umbelled ditto
———— Coeli-roſa.	Smooth-leaved ditto
Lychnis Chalcedonica.	Single flowered ſcarlet Lychnis
———— v.	
———— v. alb.	White flowered ditto
———— v. fl. pleno.	Double flowered ditto
—— Flos-cuculi.	Red flowered ditto
———— v. fl. alb.	White ditto

Lychnis

Lychnis Vifcaria.	Single flowered vifcous ditto
———— v. fl. alb.	White flowered ditto
———— v. fl pleno.	Double flowered ditto
——— Alpina.	Alpine ditto
——— dioica.	Red campion
Ceraftium perfoliatum.	Perfoliate Ceraftium
———— vulgatum.	Moufe ear chick-weed
———— vifcofum.	Clammy Ceraftium
———— femidecandrum.	Leaft ditto
———— tetrandrum.	
———— arvenfe.	Corn ditto
———— dichotomum.	Forked ditto
———— Alpiuum.	Alpine ditto
———— repens.	Creeping ditto
———— fuffruticofum.	Underfhrub ditto
———— aquaticum.	Water ditto
———— latifolium.	Broad leaved ditto
———— tomentofum.	White ditto
———— dioicum.	Spanifh ditto
Spergula arvenfis.	Corn fpurry
——— nodofa.	Knotted ditto
——— laricina.	Larix leaved ditto

DECANDRIA DECAGYNIA.

Phytolacca decandra.	Branching Phytolacca

CLASSIS XI.

DODECANDRIA MONOGYNIA.

Afarum Europeum.	Common afarabacca
——— Canadenfe.	Canadian ditto
——— Virginicum.	Sweet fcented ditto
Portulaca oleracea.	Garden purflane
———— v. fol. aur.	Golden ditto
Lythrum falicaria.	Purple fpiked willow-herb

Lythrum

| Lythrum virgatum. | Fine branched willow-herb |
| ———— hyffopifolia. | Three-leaved ditto |

DODECANDRIA DIGYNIA.

Agrimonia Eupatoria.	Common agrimony
———— odorata.	Sweet fcented ditto
———— repens.	Creeping ditto
———— agrimonoides.	Three-leaved ditto

DODECANDRIA TRIGYNIA.

Refeda luteola.	Dyer's weed
———— alba.	Upright white Refeda
———— undulata.	Wave-leaved ditto
———— lutea.	Yellow ditto
———— phyteuma.	Trifid ditto
———— odorata.	Mignonette
Euphorbia maculata.	Spotted fpurge
———— peplus.	Petty ditto
———— exigua.	Dwarf ditto
———— Lathyris.	Caper ditto
———— Portlandica.	Portland ditto
———— Paralias.	Sea ditto
———— Heliofcopia.	Wart-wort
———— pilofa.	Hairy fpurge
———— Platyphyllos.	Broad notched leaved ditto
———— Efula.	Gromwell leaved ditto
———— Cypariffias.	Cyprefs ditto
———— Myrfinites.	Glaucous ditto
———— paluftris.	Marfh ditto
———— Hiberna.	Irifh ditto
———— amygdaloides	Wood ditto

DODECANDRIA DODECAGYNIA.

| Sempervivum tectorum. | Common houfe-leek |
| ———— globiferum. | Globular ditto |

Sempervivum

Sempervivum arachnoideum.	Cobweb houfe-leek
———— hirtum.	Hairy ditto
———— montanum.	Mountain ditto
———— fediforme.	Stone-crop leaved ditto

CLASSIS XII.

ICOSANDRIA PENTAGYNIA.

Spiræa aruncus.	Goats-beard Spiræa
———— Filipendula.	Single flowered drop-wort
———— *v. fl. pleno.*	Double ditto
———— ulmaria.	Single flowered meadow fweet
———— *v. fl. pleno.*	Double ditto
———— *v. fol. variegat.*	Variegated ditto
———— lobata.	Lobe leaved Spiræa
———— trifoliata.	Three-leaved ditto

ICOSANDRIA POLYGYNIA.

Rubus faxatilis.	Stone bramble
———— Arcticus.	Dwarf ditto
———— Chamæmorus.	Mountain ditto
Fragaria Vefca.	Wood ftrawberry
———— *v. pratenfis.*	Hautboy ditto
———— *v. Chiloenfis.*	Chili ditto
———— *v. Virginiana.*	Scarlet Virginian ditto
———— *v. ananas.*	Pine ditto
———— monophylla.	Simple leaved ditto
———— fterilis.	Barren ditto
Potentilla anferina.	Wild tanfy
———— fericea.	Silky cinquefoil
———— multifida.	Multifid ditto
———— rupeftris.	Rock ditto
———— bifurca.	Bifid-leaved ditto
———— Penfylvanica.	Agrimony-leaved ditto

Potentilla

Potentilla fupina.	Trailing cinquefoil
———— recta.	Upright ditto
———— argentea.	Sattin ditto
———— hirta.	Hairy ditto
———— opaca.	Hairy Potentilla
———— verna.	Spring ditto
———— aurea.	Golden ditto
———— Aftracanica.	Aftracan ditto
———— alba.	White ditto
———— Valderia.	
———— reptans.	Common ditto
———— Monfpelienfis.	Montpelier ditto
———— Norwegica.	Norway ditto
———— grandiflora.	Great flowered ditto
———— tridentata.	Trifid leaved ditto
Tormentilla erecta.	Common Tormentil
———————— reptans.	Creeping ditto
Geum Virginianum.	American avens.
——— urbanum.	Common ditto
——— intermedium.	Wood ditto
——— Canadenfe.	Canadian ditto
——— montanum.	Mountain ditto
——— ariftatum.	Awned ditto
——— rivale.	Water ditto
——— hybridum.	Baftard ditto
——— potentilloides.	Siberian ditto
Dryas octopetala.	Mountain Dryas
Comarum paluftre.	Marfh Cinquefoil

CLASSIS XIII.

POLYANDRIA MONOGYNIA.

Actæa fpicata.	Herb Chriftopher
———— racemofa.	American ditto

N   Sanguinaria

Sanguinaria Canadenſis.	Canadian Sanguinaria.
Podophyllum peltatum.	Duck's foot
Chelidonium majus.	Common celandine
——————— laciniatum.	Cut-leaved ditto
——————— glaucium.	Yellow horned poppy
——————— corniculatum.	Red celandine
——————— hybridum.	Violet coloured ditto
Papaver hybridum.	Baſtard poppy
——————— argemone.	Rough ditto
——————— nudicaule.	Naked ſtalked ditto
——————— Rhœas.	Single red ditto
——————— v. fl. pleno.	Double ditto
——————— dubium.	Smooth ditto
——————— ſomniferuum.	Common white ditto
——————— v. ſemine nigro.	Black ſeeded ditto
——————— v. fl. pleno.	Double flowered ditto
——————— Cambricum.	Welſh ditto
——————— Orientale.	Oriental ditto
Argemone Mexicana.	Prickly Argemone, or poppy
Nymphæa lutea.	Yellow water lily
——————— alba.	White ditto
Ciſtus tuberarius.	Plantain leaved Ciſtus
——— guttatus.	Annual ſpotted flowered ditto
——— ſalicifolius.	Willow leaved annual ditto
——— Ægyptiacus.	Egyptian ditto

POLYANDRIA DIGYNIA.

Pæonia officinalis.	Common Pæony
——————— v. fl pleno rub.	Double red flowered ditto
——————— v. fl. pleno pallid.	Double pale flowered ditto
——— corallina.	Intire ditto
——— albiflora.	White flowered ditto
——— anomala.	Jagged leaved ditto
——— hybrida.	Hybrid ditto
——— tenuifolia.	Slender-leaved ditto

POLYANDRIA

POLYANDRIA TRIGYNIA.

Delphinum Ajacis.	Single upright larkfpur
———— v. fl. pleno.	Double ditto
———— v. pumil.	Dwarf ditto
———— Confolida.	Branching ditto
———— grandiflorum.	Single great flowered ditto
———— v. fl. pleno.	Double ditto
———— intermedium.	Palmated ditto
———— elatum.	Common bee ditto
———— exaltatum.	American ditto
———— Puniceum.	Scarlet ditto
Aconitum Napellus.	Monks hood
———— Lycoctonum.	Great yellow wolfs-bane
———— Japonicum.	Japan ditto
———— Anthora.	Wholefome ditto
———— variegatum.	Variegated ditto
———— album.	White ditto
———— commarum.	Purple ditto
———— uncinatum.	American ditto
———— volubile.	Twining ditto

POLYANDRIA TETRAGYNIA.

Cimicifuga fœtida.	Fœtid Cimicifuga

POLYANDRIA PENTAGYNIA.

Aquilegia vulgaris.	Common Columbine
———— v. fl. pleno.	Double flowered ditto
———— v. fl. rub.	Double red ditto
———— v. fl. alb.	Double white ditto
———— alpina.	Alpine ditto
———— Canadenfis.	Canadian Columbine
———— Sibirica.	Siberian ditto
Nigella Damafcena.	Common fennel flower
———— fativa.	Small ditto

Nigella

Nigella arvenſis.	Field fennel flower
—— Hiſpanica.	Spaniſh ditto
—— Orientalis.	Yellow ditto

POLYANDRIA HEXAGYNIA.

Stratiotes aloides.	Water aloe

POLYANDRIA POLYGYNIA.

Anemone Hepatica.	Single red Hepatica
———— *v. fl. pleno.*	Double ditto
———— *v. fl. cærul. ſimp.*	Single blue ditto
———— *v fl pleno.*	Double ditto
———— *v. fl. alb.*	White ditto
——— patens.	Wooly leaved ditto
——— Baldenſis.	Creeping rooted ditto
——— vernalis.	Vernal ditto
——— pulſatilla.	Paſque-flower ditto
——— pratenſis.	Meadow Anemone
——— alpina.	Alpine ditto
——— coronaria.	Narrow leaved garden ditto
———— *v. fl. pleno.*	Narrow leaved double ditto
——— hortenſis *v. fl. ſimp.*	Broad-leaved ditto
———— *v. fl. pleno.*	Broad leaved double ditto
——— ſylveſtris.	Large white flowered wood ditto
——— Virginiana.	Virginian ditto
——— Penſylvanica.	Penſylvanian ditto
——— nemoroſa.	Wood ditto
———— *v. fl. pleno.*	Double flowered ditto
——— appennina.	Mountain wood ditto
——— ranunculoides.	Yellow wood ditto
——— thaliĉtroides.	Meadow rue leaved ditto
Atragene alpina.	Alpine Atragene
Clematis reĉta.	Upright virgins bower
——— ochroleuca.	Yellow flowered ditto
——— integrifolia.	Entire leaved ditto

Thaliĉtrum

Thalictrum alpinum.	Alpine meadow rue
———— fœtidum.	Fœtid ditto
———— Cornuti.	Canadian ditto
———— dioicum.	Dioecious ditto
———— minus.	Small ditto
———— majus.	Great ditto
———— medium.	Middle ditto
———— Sibiricum.	Siberian ditto
———— rugosum.	Rough ditto
———— angustifolium.	Narrow-leaved ditto
———— nigricans.	Black ditto
———— flavum.	Shining leaved ditto
———— simplex.	Simple stalked ditto
———— lucidum.	Common ditto
———— aquilegifolium.	Feathered columbine
———— contortum.	Twisted ditto
———— speciosum.	Glaucous leaved ditto
Adonis autumnalis.	Common Flos Adonis
———— vernalis.	Perrenial, or spring ditto
Ranunculus Flammula.	Small spearwort crowfoot
———— reptans.	Narrow leaved ditto
———— Lingua.	Great ditto
———— gramineus.	Grass leaved ditto
———— Pyrenæus.	Pyrenean ditto
———— parnassifolius.	Parnassia-leaved ditto
———— amplexicaulis.	Plantain-leaved ditto
———— Ficaria.	Common pilewort
———— vi fl. pleno.	Double flowered ditto
———— Cassubicus.	Various leaved ditto
———— auricomus.	Wood crow-foot, or goldilocks
———— sceleratus.	Marsh ditto
———— aconitifolius.	Aconite leaved crow-foot
———— v. fl. pleno.	Double aconite ditto
———— platanifolius.	Platanus leaved ditto
———— Illyricus.	Illyrian ditto

Ranunculus

Ranunculus Penfylvanicus.	Penfylvanian crow-foot
———— Afiaticus.	Perfian ditto, or garden Ranunculus
———— rutæfolius.	Rue leaved ditto
———— nivalis.	Alpine yellow ditto
———— alpeftris.	Alpine ditto
———— bulbofus.	Bulbous ditto
———— *v fl. pleno.*	Double bulbous ditto
———— repens.	Creeping ditto
———— *v. fl. pleno.*	Double flowered ditto
———— acris.	Upright ditto
———— *v. fl. pleno.*	Double upright ditto
———— hirfutus.	Pale leaved ditto
———— lanuginofus.	Broad-leaved ditto
———— arvenfis.	Corn ditto
———— muricatus.	Spreading prickly capfuled ditto
———— parviflorus.	Small flowered ditto
———— hederaceus.	Ivy ditto
———— aquatilis.	Various leaved water ditto
———— *v. circinata.*	Fennel leaved ditto
Trollius Europeus.	European globe flower
———— Afiaticus.	Afiatic ditto
Ifopyrum fumarioides.	Fumatory-leaved Ifopyrum
Helleborus hyemalis.	Winter Hellebore
———— niger.	Black ditto, or Chriftmas rofe
———— viridis.	Green ditto
———— fœtidus.	Bears foot
———— lividus.	Great three leaved black hellebore
———— trifolius.	Small three leaved ditto
Caltha paluftris.	Marfh marygold
———— *v. fl. pleno.*	Double ditto
Hydraftris Canadenfis.	Yellow root

CLASSIS

CLASSIS XIV.

DIDYNAMIA GYMNOSPERMIA.

Ajuga Orientalis.	Oriental bugle
———— alpina.	Mountain ditto
———— Genevenfis.	Flefh coloured ditto
———— reptans.	Common ditto
———————— *v. fl. alb.*	White flowered ditto
Teucrium campanulatum.	Small flowered ditto
———— Pyrenaicum.	Pyrenean ditto
———— Botrys.	Cut leaved annual ditto
———— Chamæpitys.	Ground-pine ditto
———— multiflorum.	Many flowered ditto
———— Regium.	Box leaved ditto
———— Canadenfe.	Nettle-leaved ditto
———— hircanicum.	Betony-leaved ditto
———— Scorodonia.	Wood ditto
———— Scordium.	Water ditto
———— Chamædrys.	Common ditto
———— lucidum.	Shining ditto
———— montanum.	Dwarf ditto
———— polium.	White mountain ditto
———————— *v. fupinum.*	Sea ditto
Satureja montana.	Winter favory
———— hortenfis.	Summer ditto
Hyffopus officinalis.	Common hyffop
———— Lophanthus.	Mint-leaved ditto
———— Nepetoides.	Square ftalked ditto
Nepeta Cataria.	Common catmint
———— violacea.	Violet coloured ditto
———— incana.	Hoary ditto
———— cærulea.	Blue flowered ditto
———— Nepetella.	Small ditto
———— nuda.	Spanifh ditto

Nepeta

Nepeta Italica.	Italian catmint
———— ucranica.	Spanish ditto
———— botryoides.	Cut-leaved ditto
Sideritis perfoliata.	Perfoliate iron-wort
——— montana.	Mountain ditto
——— elegans.	Dark flowered ditto
——— hyssopifolia.	Hyssop-leaved ditto
——— scordioides.	Crenated ditto
——— hirsuta.	Hairy ditto
Elscholzia cristata.	Crested Elscholzia
Mentha villosa. *v.* 1*ma. Sole.*	Long-leaved horse mint
———————— *v.* 2*da. ditto.*	Common ditto
——— sylvestris *ditto*	Strong scented ditto
——— ——— *v. fol. varieg.*	Variegated ditto
——— rotundifolia *Sole.*	Round-leaved ditto
——— viridis *ditto.*	Spear ditto
——— palustris *ditto.*	Marsh ditto
——— piperita *v. officin. ditto.*	True pepper ditto
——————— *v. sylvestris ditto.*	Wild pepper ditto
——————— *v. vulg. ditto.*	Common ditto
——— odorata *ditto.*	Burgamot ditto
——— aquatica. *v. minor Sole.*	Small water ditto
——— ——— *v. major ditto.*	Great ditto
——————— *v. tomentosa.*	Hoary ditto
——— arvensis *Sole.*	Corn ditto
——— præcox *ditto.*	Early flowered ditto
——— agrestis *ditto.*	Field ditto
——— gentilis *ditto.*	Cardiac ditto}
——— gracilis *ditto.*	Slender ditto
——— pratensis *ditto.*	Meadow ditto
——— rubra *ditto.*	Common red mint
——— variegata *ditto.*	Window mint
——— rivalis *ditto.*	Brook mint
——— ——— *v. ditto.*	Variety of
——— ——— *v. ditto.*	Ditto
——— sativa *ditto.*	Tall red mint

Mentha

Mentha paludofa	*Sole.*	Fen mint
——— pulegium	*ditto.*	Pennyroyal
——— multiflora.		Many flowered ditto
——— crifpa.		Curled ditto
——— cervina		Hyffop-leaved ditto
——— incana.		Hoary ditto
Glecoma hederacea		Ground Ivy
Lamium Orvala.		Balm-leaved archangel
——— lævigatum.		Smooth ditto
——— garganicum.		Woolly ditto
——— rugofum.		Purple wrinkled ditto
——— — *v. fl. alb.*		White ditto
——— maculatum.		Spotted ditto
——— album.		White ditto, or dead nettle
——— purpureum.		Purple ditto
——— molle.		Pellitory leaved ditto
——— amplexicaule.		Porfoliate ditto
——— diffeɛtum.		Cut-leaved ditto
Galeopis Ladanum.		Red dead nettle
——— Tetrahit.		Common ditto
——— cannabina.		Hemp ditto
——— villofa.		Hairy ditto
———Galeobdolon.		Yellow ditto
Betonica ftriɛta.		Danifh betony.
——— officinalis.		Wood ditto
——— incana.		Hoary ditto
——— Orientalis.		Oriental ditto
——— hirfuta.		Hairy ditto
Stachys fylvatica.		Wood Stachys
——— paluftris,		Marfh ditto
——— alpina.		Alpine ditto
——— circinata.		Blunt leaved ditto
——— Germanica.		Bafe horehound
——— lanata.		Woolly Stachys.
——— Cretica.		Cretan ditto

O   Stachys

Stachys hirta.	Procumbent Stachys
——— recta.	Upright ditto
——— annua.	White annual ditto
——— arvensis.	Corn ditto
Ballota nigra.	Black horehound
———v. fl. alb.	White flowered ditto
——— lanata.	Woolly ditto
Marrubium 'Alyffon.	Plaited-leaved white horehound
——— peregrinum.	
——— candidiffimum.	Woolley white ditto
——— fupinum.	Procumbent white ditto
——— vulgare.	Common white ditto
——— Hifpanicum.	Spanifh white ditto
Leonurus Cardiaca.	Common mother-wort
——— Marrubiaftrum.	Small flowered ditto
——— Tartaricus.	Tartarian ditto
——— Sibiricus.	Siberian ditto
Phlomis Herba-venti.	Rough-leaved Phlomis
——— tuberofus.	Tuberous ditto
Moluccella lævis.	Smooth Molucca balm
——— fpinofa.	Prickly ditto
Clinopodium vulgare.	Wild bafil
Origanum Creticum.	Cretan Origanum
——— heracleoticum.	Winter fweet majoram
——— vulgare.	Common ditto
——— Onites.	Pot ditto
——— Majorana.	Sweet, or knotted marjoram
Thymus Serpyllum.	Common fmooth mother of thyme
——— v citri-odor.	Lemon thyme
——— v. villofum,	Hoary mother of thyme
——— v. hirfutum.	Hairy ditto
——— vulgaris.	Broad-leaved garden ditto
——— Zygis.	Linear-leaved ditto
——— Acinos.	Corn ditto
——— Patavinus.	Great flowered ditto

Thymus

Thymus Virginicus.	Virginian, or favory ditto
Meliſſa officinalis.	Common balm
—————v. varieg.	Variegated ditto
——grandiflora.	Great flowered ditto
————v. fol. varieg.	Variegated ditto
——— Calamintha.	Mountain ditto
———- Nepeta.	Field ditto
Dracocephalum Virginianum	Virginian dragon's head
——————— denticulatum.	Denticulated ditto
——————— peregrinum.	Prickly-leaved ditto
——————— Austriacum.	Austrian ditto
——————— Ruyſchianum.	Hyſſop-leaved ditto
——————— Sibiricum.	Siberian ditto
——————— Moldavicum.	Moldavian ditto
——————— caneſcens.	Hoary ditto
——————— peltatum.	Willow-leaved ditto
——————— grandiflorum.	Great flowered ditto
——————— nutans.	Nodding ditto
——————— thymiflorum.	Small flowered ditto
Melittis Meliſſophyllum.	Common baſtard balm
——— grandiflorum.	Great flowered ditto
Ocymum Baſilicum.	Common ſweet baſil
Scutellaria albida.	Hairy ſkull-cap
——— Alpina.	Alpine ditto
——— lupulina.	Great flowered ditto
——— lateriflora.	Virginian ditto
——— galericulata.	Common ditto
——— minor.	Leaſt ditto
——— integrifolia.	Entire leaved ditto
——— peregrina.	Florentine ditto
——— altiſſima.	Tall ditto
Prunella vulgaris.	Common ſelf-heal
———————v. fl. alb.	White flowered ditto
——— grandiflora.	Great flowered ditto
——— intermedia.	
——— laciniata.	Jagged-leaved ditto

O 2   Didynamia

DIDYNAMIA ANGIOSPERMIA.

Bartſia viſcoſa.	Marſh Bartſia
——— Alpina.	Mountain ditto
Rhinanthus Criſta-Galli.	Yellow rattle
Euphraſia officinalis.	Common eye-bright
——— Odontites.	Red ditto
Melampyrum pratenſe.	Meadow cow wheat
——— ——— ſylvaticum.	Wood ditto
Pedicularis paluſtris.	Marſh louſe wort.
——— ſylvatica.	Common ditto
Chelone glabra.	White Chelone
——— obliqua.	Red ditto
Antirrhinum Cymbalaria.	Ivy-leaved toad-flax
———Elatine.	Sharp pointed ditto
——— ſpurium	Round leaved ditto
——— medium.	Two coloured ditto
——— cirrhoſum.	Tendrill'd ditto
——— triphyllum.	Three-leaved ditto
——— purpureum.	Purple ditto
——— verſicolor,	Spiked flowered ditto
——— repens	Creeping ditto
——— Monſpeſulanum.	Montpelier ditto
——— ſparteum.	Branching ditto
——— bipunctatum.	Dotted flowered ditto
——— arvenſe.	Blue flowered corn ditto
——— v. fl. lut.	Yellow flowered ditto
——— viſcoſum.	Clammy ſnap-dragon
——— minus.	Leſſer toad-flax
——— geniſtifolium.	Broom-leaved ditto
——— Chalepenſe.	White-flowered ditto
——— Linaria.	Common yellow ditto
——— v. peloria.	
——— majus.	Common ſnap-dragon
——— v. fl. ruberrim.	Scarlet-flowered ditto

Antirrhinum

Antirrhinum majus *v. fl. alb.*	White-flowered ditto
—————— *v. fl. pleno.*	Double-flowered ditto
—————— Orontium.	Calf's snout
—————— Afarina.	Heart-leaved toad-flax
Scrophularia nodofa.	Knobby rooted fig-wort.
—————— aquatica.	Water ditto
—————— auriculata.	Ear-leaved ditto
—————— Scorodonia.	Balm-leaved ditto
—————— Altaica.	White-flowered ditto
—————— betonicifolia.	Betony-leaved ditto
—————— Orientalis.	Hemp-leaved ditto
—————— vernalis.	Yellow ditto
—————— fambucifolia.	Elder-leaved ditto
—————— cananina.	Cut-leaved ditto
—————— lucida.	Shining-leaved ditto
—————— peregrina.	Nettle-leaved ditto
Celfia Cretica.	Great flowered Celfia
Digitalis purpurea.	Purple fox glove
—————— *v. fl. albo.*	White flowered ditto
—————— minor.	
—————— lutea.	Small yellow fox-glove
—————— ambigua.	Greater yellow ditto
—————— ferruginea.	Iron coloured ditto
—————— ciliaris.	Ciliated foxglove
Erinus alpinus.	Alpine Erinus
Sibthorpia Europea.	Cornifh Sibthorpia
Limofella aquatica.	Mudwort
Dodartia Orientalis.	Oriental Dodartia
Pentstemon pubefcens.	Hairy Pentftemon
—————— lævigata.	Smooth ditto
Mimulus ringens	Oblong-leaved monkey-flower.
—————— alatus.	Oval-leaved ditto.
Acanthus mollis.	Smooth bears birch
—————— *v. nigr.*	Portuguefe ditto
—————— fpinofus.	Prickly ditto

CLASSIS

CLASSIS XV.

TETRADYNAMIA SILICULOSA.

Myagrum perenne.	Perrenial gold of pleafure
———— Orientale.	Oriental ditto
———— rugofum.	Rough ditto
———— perfoliatum.	Perfoliate ditto
———— fativum.	Cultivated ditto
———— paniculatum.	Panicled ditto
———— faxatile.	Rock ditto
Vella annua.	Annual Vella
Draba verna.	Common whitlow-grafs
—— aizoides.	Hairy leaved Alpine ditto
—— Pyrenaica.	
—— muralis.	Wall ditto
—— hirta.	Hairy ditto
—— incana.	Hoary ditto
Lepidium perfoliatum.	Various leaved pepper-wort
———— alpinum.	Alpine ditto
———— petræum.	Rock ditto
———— fpinofum.	Prickly ditto
———— fativum.	Garden, or common crefs
———— lyratum.	
———— latifolium.	Broad leaved ditto
———— didymum.	Procumbent ditto
———— ruderale.	Narrow leaved ditto
———— Iberis.	Bufhy ditto
Thlafpi arvenfe.	Hairy ditto
———— faxatile.	Rock ditto
———— hirtum.	Fied baftard crefs
———— campeftre.	Wild ditto, or mithridate muf- tard
———— montanum.	Mountain ditto
———— perfoliatum.	Pefoliate baftard-crefs

Thlafpi

Thlaſpi Burſa-paſtoris.	Shepherd's purſe
—— Ceratocarpon.	Siberian baſtard creſs.
Cochlearia officinalis.	Common ſcurvy graſs
—— Danica.	Daniſh ditto
—— Anglica.	Sea ditto
—— Coronopus.	Wild ditto, or ſwines-creſs
—— Armoracia.	Horſe-radiſh
—— glaſtifolia.	Wood leaved ſcurvy-graſs
—— Draba.	
Iberis ſempervirens.	Narrow leaved evergreen candy-tuft
—— ſaxatilis.	Rock ditto
—— umbellata.	Purple ditto
—————— v.	Normandy ditto
—— amara.	White candy tuft
—— nudicaulis..	Naked ſtalked ditto
—— pinnata.	Winged ditto
Alyſſum halimifolium.	Sweet ſcented madwort
—— ſaxatile.	Rock ditto
—— incanum.	Hoary ditto
—— minimum.	Small ditto
—— montanum.	Mountain ditto
—— clypeatum.	Upright ditto
—— ſinuatum.	Sinuated ditto
—— Creticum.	Cretan ditto
—— utriculatum.	Bladder ditto
—— deltoideum.	Purple ditto
Peltaria alliacea.	Garlick ſcented Peltaria
Biſcutella articulata.	Ear podded Biſcutella
—— apula.	Spear leaved ditto
—— lævigata.	Smooth ditto
Lunaria rediviva.	Perennial honeſty
—— annua.	Annual ditto

TETDADYNAMIA

TETRADYNAMIA SILIQUOSA.

Dentaria bulbifera.	Coral-wort
———— pentaphyllos.	Five leaved toothwort
Cardamine bellidifolia.	Daisy leaved ladies smock
———— asarifolia.	Heart leaved ditto
———— nivalis.	
———— petræa.	Mountain ditto
———— trifolia.	Three leaved ditto
———— impatiens.	Impatient ditto
———— hirsuta.	Hairy ditto
———— pratensis. *v. fl. simp.*	Common ditto
———— *v. fl. pleno.*	Double flowered ditto
———— amara.	Bitter ladies smock
Sisymbrium Nasturtium.	Common water cress
———— sylvestre.	Water rocket
———— amphibium.	Water radish
———— terrestre.	Various leaved ditto
———— Pyrenaicum.	Pyrenean Sisymbrium
———— polyceratium.	Dandelion leaved ditto
———— murale.	Wall ditto
———— monense.	Procumbent ditto
———— Sophia.	Flix weed
———— Irio.	Broad leaved hedge-mustard
———— Læselii.	Hairy Sisymbrium
———— Columnæ.	Hoary leaved ditto
———— Supinum.	Trailing ditto
———— Pannonicum.	Hungarian ditto
———— Barbarea.	
———— strictissimum.	Spear leaved ditto
Erysimum officinale.	Common hedge mustard
———— Barbarea. *v. fl. simp.*	Winter cress
———— *v. fl. plen.*	Double flowered ditto
———— Alliaria.	Sauce-alone
———— repandum.	Small flowered hedge-mustard

Erysimum

Eryfimum cheiranthoides.	Wormfeed
—— hieracifolium.	Saw-leaved ditto
Cheiranthus eryfimoides.	Wild ftock
—— alpinus.	Alpine ditto
—— Cheiri *v. fl. lut. fimp.*	Single yellow wall-flower
—— *v. fl. plen.*	Double yellow ditto
—— *v. fl. coccin. fimp.*	Single bloody ditto
—— *v. fl. plen.*	Double bloody ditto
—— maritimus *v. fl. viol.*	Dwarf annual ftock
—— *v. fl. alb.*	White flowered ditto
—— fruticulofus.	Willow-leaved ditto
—— incanus.	Queen ftock gillyflower
—— *v. fl. coccin.*	Bromptons ftock ditto
—— *v. fl. alb.*	White ftock ditto
—— *v. glab.*	Wall flower leaved ditto
—— annuus.	Ten weeks ftock ditto
—— triftis.	Dark flowered ditto
—— tricufpidatus.	Trifid ditto
—— finuatus.	Greater fea ftock
Hefperis triftis.	Night fmelling rocket
—— matronalis *v. fl. alb. fimp.*	Single white ditto
—— *v. fl. plen.*	Double flowered ditto
—— *v. fl. purp. fimp.*	Single purple ditto
—— *v. fl. plen.*	Double ditto
—— Africana.	African ditto
—— verna.	Early flowering ditto
Arabis alpina.	Alpine wall crefs
—— lucida.	Shining-leaved ditto
—— thaliana.	Common ditto
—— bellidifolia.	Daify-leaved ditto
—— hifpida.	Rough ditto
—— turrita.	Tower ditto
Turritis glabra.	Smooth tower muftard
—— hirfuta.	Hairy ditto
Braffica Orientalis.	Perfoliate cabbage

P   Braffica

Braffica campeftris.	Field cabbage
—— Napus.	Wild ditto
—— Rapa.	Turnep
—— oleracea *v. alb.*	White cabbage
—— *v. rub.*	Red ditto
—— *v. fabad.*	Savoy ditto
—— *v. fabellic. virid.*	Borecole ditto
—— *v. purp.*	Red borecole ditto
—— *v. botrytis.*	Cauliflower
—— *v. Italica alba.*	White cauliflower broccoli
—— *v. Italica purp.*	Purple ditto
—— *v.*	Brofels fprout
—— *v. Napo-braffica*	Turnep rooted cabbage
—— muralis.	Wall ditto
—— Eruca.	Hairy ftalked ditto
—— veficaria.	Hairy podded ditto
Sinapis arvenfis.	Charlock
—— Orientalis.	Oriental muftard
—— alba.	White ditto
—— nigra.	Common black ditto
—— pubefcens.	Pubefcent ditto
—— juncea.	Fine-leaved ditto
—— erucoides.	
—— lævigata.	Smooth ditto
—— incana.	Hoary ditto
Raphanus fativus *v. purp.*	Common garden radifh
—— *v. pallid.*	
—— *v. Rapa.*	Turnep rooted ditto
—— *v. niger.*	Black Spanifh ditto
—— Raphaniftrum.	Wild charlock
—— tenellus.	Small radifh
Bunias Erucago.	Prickly podded Bunias
—— Orientalis.	Oriental ditto
—— Cakile.	Sea ditto
—— Balearica.	Minorca ditto
Ifatis tinctoria.	Dyers weed

Crambe

Crambe maritima.	Sea colewort
———— Hifpanica.	Spanifh ditto
———— Tartarica.	Tartarian ditto

CLASSIS XVI.

MONADELPHIA PENTANDRIA.

Erodium Romanum.	Roman cranes-bill
———— cicutarium *v. fl. plen.*	Hemlock leaved ditto
———— *v. fl. alb.*	White flowered ditto
———— bimaculatum.	Two fpotted ditto
———— mofchatum.	Mufk ditto
———— malacoides.	Mallow leaved ditto
———— maritimum.	Sea ditto
———— gruinum.	Broad leaved annual ditto
———— ciconium.	Long bearded ditto
———— trilobatum.	Three lobed ditto
———— chamædryoides.	Dwarf ditto

MONADELPHIA DECANDRIA.

Geranium tuberofum.	Tuberous rooted cranes-bill
———— macrorhizum.	Long rooted ditto
———— phæum.	Dwarf flowered ditto
———— reflexum.	Purple flowered ditto
———— lividum.	Wrinkled leaved ditto
———— nodofum.	Knotty ditto
———— ftriatum.	Streaked ditto
———— fylvaticum.	Wood ditto
———— paluftre.	Marfh ditto
———— pratenfe *v. fl. cærul.*	Meadow ditto
———— *v. fl. alb.*	White flowered meadow ditto
———— argenteum.	Silvery leaved ditto
———— maculatum.	Spotted ditto
———— Pyrenaicum.	Mountain ditto

Geranium aconitifolium.	Aconite-leaved ditto
—————— angulatum.	Angular ſtalked ditto
—————— Bohemicum.	Bohemian ditto
—————— Robertianum *v. fl.*	
rub.	Herb-robert
———————— *v. fl. alb.*	White flowered ditto
—————————— *v. ſaxatile.*	Rock ditto
—————— lucidum.	Shining ditto
—————— molle *v. fl. purp.*	Doves foot ditto
———————— *v. fl. alb.*	White flowered ditto
—————— Carolinianum.	Carolina ditto
—————— Columbinum.	Long ſtalked ditto
—————— diſſectum.	Jagged ditto
—————— rotundifolium.	Round leaved ditto
—————— puſillum.	Small flowered ditto
———— Sibiricum.	Siberian ditto
—————— ſanguineum.	Bloody ditto
—————— Lancaſtrienſe.	Lancaſhire ditto

MONADELPHIA POLYANDRIA.

Althæa officinalis.	Common marſh mallow
—————— cannabina.	Hemp leaved ditto
—————— hirſuta.	Hairy ditto
Alcea roſea *v. fl. ſimp. purp.*	Single flowered purple holly-hock
—————— *v. fl. pleno.*	Double flowered ditto
—————— *v. fl. lut. ſimp.*	Single flowered yellow ditto
—————— *v. fl. plen.*	Double flowered ditto
————— *v. Chinenſis.*	Chineſe ditto
———— ſicifolia.	Fig-leaved ditto
Malva Americana.	American mallow
—— Peruviana.	Peruvian ditto
—— Limenſis.	Blue flowered ditto
—— Caroliniana.	Creeping ditto
—— parviflora.	Small flowered ditto

Malva

Malva rotundifolia.	Round leaved ditto
—— fylveſtris.	Common ditto
—— Mauritiana.	Ivy-leaved ditto
—— verticillata.	Whorl flowered ditto
—— criſpa.	Curled ditto
—— Alcea.	Vervain ditto
—— moſchata *v. fl. purp.*	Purple flowered muſk ditto
——————— *v. fl. alb.*	White flowered ditto
Lavatera arborea.	Tree mallow
—— Thuringiaca.	Great flowered ditto
—— Cretica.	Cretan ditto
—— trimeſtris *v. fl. roſ.*	Annual ditto
——————— *v. fl. alb.*	White flowered ditto
Hibiſcus paluſtris.	Marſh Hibiſcus
—— Trionum.	Common bladder ditto

CLASSIS XVII.

DIADELPHIA HEXANDRIA

Fumaria cucularia.	Naked ſtalked fumitory
—— nobilis.	Great flowered ditto
—— ſolida.	Solid rooted bulbous fumitory
—— cava *v. fl. albo.*	Hollow rooted bulbous white ditto
——————— *v. fl. rub.*	Hollow rooted bulbous red ditto
—— glauca.	Glaucous ditto
—— lutea.	Yellow ditto
—— capnoides.	White flowered ditto
—— officinalis.	Common ditto
—— capreolata.	Ramping ditto
—— tenuifolia.	Slender leaved ditto
—— ſpicata.	Narrow leaved ditto
—— claviculata.	Climbing ditto

Fumaria

Fumaria veſicaria.	Bladdered fumitory
——— fungoſa.	Spungy flowered ditto

DIADELPHIA OCTANDRIA.

Polygala vulgaris *v. fl. cærul.*	Blue flowered milkwort
——————— *v. fl. alb.*	White ditto
——————— *v. fl. incarnat.*	Fleſh-coloured ditto

DIADELPHIA DECANDRIA.

Ononis Antiquorum.	Tall ditto
——— ſpinoſa *v. fl. purp.*	Thorny purple reſt-harrow
——— ——— *v. fl. alb.*	Thorny white ditto
——— mitiſſima.	Cluſter-flowered annual ditto
——— alopecuroides.	Foxtail ditto
——— hircina.	Stinking ditto
——— viſcoſa.	Clammy ditto
——— Columnea.	Cluſter flowered ditto
——— rotundifolia.	Round-leaved ditto
Anthyllis tetraphylla.	Four leaved kidney vetch
——————— vulneraria *v. fl. lut.*	Yellow ditto
——————— *v. fl. coccin.*	Scarlet ditto
——————— montana.	Mountain Anthyllis
Lupinus perennis.	Perennial lupine
——— albus.	White ditto
——— varius.	Small blue ditto
——— hirſutus.	Great ditto
——— piloſus.	Roſe ditto
——— anguſtifolius.	Narrow leaved blue ditto
——— luteus.	Yellow ditto
Phaſeolus vulgaris *v. fl. alb.*	Common kidney-bean
——————— *v. fl. puniceo.*	Scarlet flowered ditto
——————— nanus.	
Glycine Apios.	Tuberous rooted Glycine
Piſum ſativum *v. majus.*	Common marrow-fat pea

Piſum

Pifum fativum *v. viride*.	Green rouncival marrow-fat pea
———— *v. edul.*	Sugar ditto
———— *v nan im.*	Dwarf ditto
—— umbellatum.	Rofe, or crown ditto
—— maritimum.	Sea ditto
—— ochrus.	Yellow-flowered ditto
Orobus lathyroides.	Upright bitter vetch
——— luteus.	Yellow ditto
——— vernus.	Spring ditto
——— tuberofus.	Tuberous ditto
——— fylvaticus.	Wood ditto
——— anguftifolius.	Narrow leaved ditto
——— niger.	Black ditto
Lathyrus Aphaca.	Yellow Lathyrus
——— Niffolia.	Crimfon ditto
——— amphicarpos.	Subterranean ditto
——— Cicera.	Flat podded ditto
——— fativus *v. fl. cærul.*	Common ditto
——— *v fl. alb.*	White flowered ditto
——— inconfpicuus.	Small flowered ditto
——— fetifolius.	Narrow leaved ditto
——— angulatus.	Angular feeded ditto
——— odoratus. *v. fl. purp.*	Purple fweet pea
——— *v. Zeylanicus*	Painted lady ditto
——— *v. fl. alb.*	White flowered ditto
——— *v. coccin.*	Scarlet ditto
——— annuus.	Two flowered yellow Lathyrus
——— Tingitanus.	Tangier ditto
——— Clymenum.	Various flowered ditto
——— hirfutus.	Hairy ditto
——— tuberofus.	Tuberous ditto
——— fylveftris.	Wild ditto
——— latifolius.	Broad-leaved ditto
——— pratenfis.	Meadow ditto
——— paluftris.	Marfh ditto
——— pififormis.	Siberian ditto

Lathyrus

Lathyrus articulatus.	Jointed podded ditto
Vicia pififormis.	Pale flowered vetch
—— fylvatica.	Common wood ditto
—— canefcens.	Hoary ditto
—— Cracca.	Tufted ditto
—— niffoliana.	Red flowered ditto
—— biennis.	Biennial ditto
—— Bengalienfis.	Bengal ditto
—— fativa.	Common tare
—— alba.	White ditto
—— anguftifolia.	Narrow-leaved vetch
—— lathyroides.	Dwarf ditto
—— lutea.	Smooth flowered yellow ditto
—— hybrida.	Hairy flowered yellow ditto
—— Panonica.	
—— lævigata.	Smooth ditto
—— peregrina.	Broad podded ditto
—— fepium.	Bufh ditto
—— Bithynica.	Rough ditto
—— Faba *v.*	Common bean
——— *v.*	Long podded ditto
——— *v.*	Windfor ditto
——— *v.*	Toker ditto
—— ferratifolia.	Sawed leaved ditto
Ervum Lens.	Common lentil
——— tetrafpermum.	Smooth tare
——— hirfutum.	Hairy ditto
——— monanthus.	One flowered tare
——— Ervilia.	Officinal ditto
Cicer arietinum.	Chick pea
Glycyrrhiza echinata.	Prickly headed liquorice
——— glabra.	Common ditto
Coronilla coronata.	Crowned Coronilla
——— Securidaca.	Hatchet ditto
——— varia.	Purple ditto
——— Cretica.	Cretan ditto

Ornithopus

Ornithopus perpufillus.	Common birds foot
———— intermedius.	
———— compreffus.	Hairy ditto
———— fcorpioides.	Purflane-leaved ditto
Hippocrepis unifiliquofa.	Single-podded horfe-fhoe vetch
———— multifiliquofa.	Many-podded ditto
———— comofa.	Tufted ditto
Scorpiurus vermiculatus.	Common caterpillar
———— muricatus.	Two flowered ditto
———— fulcatus.	Furrowed ditto
Hedyfarum canadenfe.	Canadian Hedyfarum
———— obfcurum.	Creeping rooted ditto
———— coronarium *v. fl. viol.*	French honey-fuckle
———————— *v. fl. alb.*	White flowered ditto
———— fluxuofum.	Zigzag podded ditto
———— Onobrychis.	St. Foin
———— Caput-Galli.	Cocks-head Hedyfarum
———— Crifta-Galli.	Cocks-comb ditto
Galega officinalis *v. fl. cærul.*	Officinal blue goats-rue
———————— *v. fl. alb.*	Officinal white ditto
Phaca Bœtica.	Hairy Phaca, or baftard vetch
——— Alpina.	Smooth ditto
Aftragalus alopecuroides.	Fox-tail milk vetch
———— fulcata.	Furrowed ditto
———— galegiformis.	Goats rue leaved ditto
———— Onobrichis.	Purple fpiked ditto
———— tenuifolius.	Slender leaved ditto
———— uliginofus.	Violet coloured ditto
———— virefcens.	Green flowered ditto
———— Cicer.	Bladdered ditto
———— Glycyphyllos.	Liquorice ditto
———— hamofus.	Dwarf yellow flowered ditto
———— contortuplicatus.	Wave podded ditto
———— Bœticus.	Triangular podded ditto
———— fefameus.	Starry ditto
———— Auftriacus.	Auftrian ditto

Q   Aftragalus

Aftragalus epiglottis.	Heart podded ditto
———— hypoglottis.	Purple mountain ditto
——— — Sibiricus.	Siberian ditto
———— alpinus.	Alpine ditto
———— trimeftris.	Egyptian ditto
——— — montanus.	Mountain ditto
———— uralenfis.	Silky ditto
———— Monfpeffulanus.	Montpelier ditto
———— leucophæus.	White flowered ditto
——— — campeftris.	Field ditto
——— — exfcapus.	Hairy podded ditto
———— tragacantha.	Goat's thorn ditto
Biferrula Pelecinus.	Baftard hatchet vetch
Trifolium M. cærulea.	Blue mellilot trefoil
———— M. Indica.	Indian ditto
———— M. Polonica.	Polonian ditto
———— M. officinalis.	Common ditto
———— M. Italica.	Italian ditto
———— M. Cretica.	Cretan ditto
———— M. ornithopodioides.	Bird's-foot ditto
· ——— — Lupinafter.	Baftard lupine
·——— — reflexum.	Reflexed ditto
———— hybridum.	Baftard ditto
———— repens.	Dutch Clover
——————— v.	
——————— v.	
———— alpinum.	Alpine trefoil
———— fubterraneum.	Subterranean ditto
———— lappaceum.	Burdock feeded ditto
———— rubens.	Lung-fpiked ditto
——— — medium.	Long-leaved ditto
———— pratenfe.	Purple ditto
———— alpeftre.	Oval-fpiked ditto
——— — incarnatum.	Flefh coloured ditto
———— involucratum.	Striped flowered ditto
———— ochroleucum.	Pallid ditto

Trifolium

Trifolium angustifolium.	Narrow leaved trefoil
—————— arvense.	Hare's-foot ditto
————— stellatum.	Star ditto
————— maritimum	Sea ditto
————— clypeatum.	Oriental ditto
————— scabrum.	Rough ditto
————— glomeratum.	Round-headed ditto
————— striatum·	Knotted ditto
————— spumosum.	Bladdered ditto
————— tomentosum.	Woolly ditto
————— fragiferum	Strawberry ditto
————— montanum.	Mountain ditto
————— agrarium.	Hop ditto
————— spadiceum.	Pale-flowered ditto
————— procumbens.	Procumbent ditto
————— suffocatum.	Dwarf ditto
Lotus maritimus.	Sea bird's foot trefoil
——— siliculosus.	Square podded ditto
——— tetragonolobus.	Winged pea
——————— v. lut.	Yellow flowered ditto
——— angustissimus.	
——— ornithopodioides.	Claw-podded trefoil
——— rectus.	Upright flowered ditto
——— corniculatus.	Common ditto
————————— v. villosus.	Hairy ditto
Trigonella platycarpos.	Round-leaved fenugreek
————— spinosa.	Thorny ditto
————— polycerata.	Broad-leaved, or Spanish ditto
————— Monspeliaca.	Trailing ditto
————— Fœnum-græcum.	Common ditto
Medicago radiata.	Ray podded ditto
————— circinata.	Kidney-podded medick
————— sativa.	Lucern
————— falcata.	Yellow mudick
————— lupulina.	Black ditto
————— polymorpha.	Variable ditto

Medicago polymorpha.

v. orbicularis.	Flat-podded medick
———— *v. fcutellata .*	Snail ditto
———— *v. tornata.*	Smooth-podded ditto
———— *v. turbinata.*	Turban ditto
———— *v. intertexta.*	Hedgehog ditto
———— *v. Arabica.*	Heart ditto
———— *v. coronata.*	
———— *v. ciliaris.*	
———— *v. rigidula.*	Thorny-podded ditto
———— *v. muricata.*	Prickly ditto
———— *v. laciniata..*	Cut-leaved ditto

CLASSIS XVIII.

POLYADELPHIA. ICOSANDRIA.

Hypericum calycinum.	Great flowered St. John's-wort
——— Androfœmum.	Common tutfan
——— hircinum *v. majus.*	Common ftinking St. John's wort.
——— *v. minus.*	Small ditto
——— quadrangulum.	St. Peter's wort
——— perforatum.	Perforated St. John's wort
——— humifufum.	Trailing ditto
——— montanum.	Mountain ditto
——— hirfutum.	Hairy ditto
——— elodes.	Marfh ditto
——— pulchrum.	Upright ditto
——— dubium.	Doubtful ditto
——— elatum.	Tall ditto
——— Olympicum.	Olympian ditto
——— proliferum.	Proliferous ditto
——— Pyramidatum.	Pyramidal ditto

CLASSIS

CLASSIS XIX.

SYNGENESIA POLYGAMIA ÆQUALIS.

Geropogon glabrum.	Smooth old man's beard
Tragopogon pratenfis.	Yellow goat's beard
———— majus.	Great ditto
———— porifolius.	Purple ditto
———— crocifolius.	Crocus-leaved ditto
———— picroides.	Prickly-cupped ditto
———— afper.	Rough ditto
Scorzonera Hifpanica.	Garden Scorzonera
———— graminifolia.	Grafs-leaved vipers grafs
———— anguftifolia.	Narrow-leaved ditto
———— laciniata.	Cut-leaved ditto
———— Tingitana.	Poppy-leaved ditto
———— Picroides.	Various-leaved ditto
Picris echioides.	Rough Picris, or ox tongue
——— hieracioides.	Yellow fuccory
Sonchus paluftris.	Marfh fow-thiftle
——— arvenfis.	Corn ditto
——— oleraceus *v. lævis*.	Smooth-leaved common ditto
———— *v. afper*.	Rough ditto
——— alpinus.	Alpine ditto
——— Sibiricus.	Willow-leaved ditto
——— Canadenfis.	Canadian ditto
Lactuca quercina.	Oak leaved Lactuca
——— intybacea.	
——— fativa *v.*	Common cabbage lettuce
———— *v.*	Brown Dutch ditto
———— *v.*	White cofs ditto
———— *v.*	Green cofs ditto
———— *v.*	Spotted
———— *v.*	Brown ditto

Lactuca

Lactuca scariola.	Prickly cabbage lettuce
———— virosa.	Strong-scented ditto
———— saligna.	Least ditto
———— perennis.	Perrenial ditto
Prenanthus viminea.	Twiggy Prenanthus
———— purpurea.	Purple ditto
———— muralis.	Wall ditto
Leontodon Taraxacum.	Common dandelion
———— aureum.	Golden ditto
———— palustre.	Marsh ditto
———— alpinum.	Alpine ditto
———— hispidum.	Rough ditto
———— hirtum.	Hairy ditto
———— autumnale *v. glab.*	Smooth autumnal ditto
———————— *v. hispid.*	Rough ditto
Hieracium incanum.	Hoary hawk-weed
———— alpinum.	Alpine ditto
———— Pilosella.	Mouse-ear ditto
———— albidum.	White ditto
———— dubium.	Creeping ditto
———— Auricula.	Narrow-leaved ditto
———— cymosum.	Small-flowered ditto
———— aurantiacum.	Orange-flowered ditto
———— porrifolium.	Leek-leaved ditto
———— murorum.	Wall ditto
———— paludosum.	Marsh ditto
———— cerinthoides.	Honeywort ditto
———— amplexicaule.	Heart-leaved ditto
———— pyrenaicum.	Pyrenean ditto
———— undulatum.	Wave-leaved ditto
———— molle.	Soft-leaved ditto
———— villosum.	Villous ditto
———— Kalmii.	Kalm's ditto
———— Sprengerianum.	Branched ditto
———— spicatum.	Hairy ditto
———— sabaudum.	Shrubby ditto

Hieracium

Hieracium v. *fol. macul.*	Spotted shrubby hawk-weed
———— umbellatum.	Umbelled ditto
———— sylvaticum.	Wood ditto
Crepis barbata.	Spanish Crepis
———— alpina.	Alpine ditto
———— rubra.	Purple ditto
———— foetida.	Stinking ditto
———— aspera.	Rough ditto
———— ragadioloides.	Intire leaved ditto
———— Sibirica.	Siberian ditto
———— tectorum.	Smooth ditto
———— biennis.	Biennial ditto.
———— Dioscoridis.	Dioscorides's ditto
———— pulchra.	Small-flowered ditto
———— coronopifolia.	
Andryala integrifolia.	Hoary Andryala
———— lanata.	Woolly ditto
Hyoseris foetida.	Stinking Hyoseris
———— radiata.	Starry ditto
———— lucida.	Shining ditto
———— minima.	Least ditto
———— hedypnois.	Branching ditto
———— rhagadioides.	Nipple-wort ditto
———— Cretica.	Cretan ditto
Seriola Æthenensis.	Rough Seriola
Hypochæris maculata.	Spotted Hypochæris
———— glabra.	Smooth ditto
———— radicata.	Long rooted ditto
Lapsana communis.	Common nipplewort
———— Zacintha.	Warted ditto
———— stellata.	Starry ditto
———— rhagadiolus.	Heart-leaved ditto
Catananche cærulea.	Blue Catananche
———— lutea.	Yellow ditto
Cichorium Intybus.	Wild endive, or succory
———— Endivia v. *latifol.*	Broad-leaved endive

Cichorium

Cichorum Endivia *v. crisp. virid.*	Green curled endive
———————— *v. crisp. alb.*	White curled ditto
———— spinosum.	Prickly ditto
Scolymus maculatus.	Annual golden thistle
Arctium Lappa.	Common burdock
Serratula tinctoria *v. fl. purp.*	Common saw wort
———————— *v. fl. alb.*	White flowered ditto
——— alpina.	Mountain ditto
——— præalta.	Tall ditto
——— spicata.	Spiked ditto
Carduus arvensis.	Way thistle
——— laceolatus.	Spear ditto
——— Monspesulanus.	Montpelier ditto
——— nutans.	Musk ditto
——— speciosus.	
——— crispus.	Curled ditto
——— tenuiflorus.	Slender flowered ditto
——— palustris *v. fl. rub.*	Common marsh ditto
———————— *v. fl. alb.*	White flowered ditto
——— pycnocephalus.	Italian ditto
——— canus.	Hoary ditto
——— defloratus.	Various leaved thistle
——— marianus.	Milk ditto
——— Syriacus.	Syrian ditto
——— eriophorus.	Woolly headed ditto
——— heterophyllus.	Various leaved ditto
——— helenioides.	Melancholy ditto
——— Tartaricus.	Tartarian ditto
——— ciliatus.	Fringed cuped ditto
——— pratensis.	Meadow ditto
——— acaulis.	Dwarf ditto
Cnicus oleraceus.	Pale-flowered Cnicus
——— Centauroides.	Artichoke leaved ditto
——— cernuus.	Siberian ditto
Onopordum acanthium.	Cotton thistle
——— Illyricum.	Illyrian Onopordum

Onopordum

Onopordum Arabicum.	Arabian Onopordum
———— acaulon.	Dwarf ditto
Cynara Scolymus *v. aculeat.*	French artichoke
————————— *v. non-aculeat.*	Globe ditto
———— Cardunculus.	Cardoon ditto
Carlina acaulis.	Dwarf Carlina
———— vulgaris.	Common ditto
Carthamus tinctorius.	Baftard faffron
———— lanatus.	Woolly Carthamus
———— Creticus.	Cretan ditto
Bidens tripartita.	Water Hemp
———— cernua.	Nodding Bidens
———— pilofa.	Hairy ditto
Cacalia Saracenica.	Creeping rooted Cacalia
———— haftata.	Spear-leaved ditto
———— fauveolens.	Sweet fcented ditto
———— alpina.	Alpine ditto
Eupatorium feffilifolium.	Seffile-leaved Eupatorium
———— trifoliatum.	Three leaved ditto
———— cannabinum.	Hemp agrimony
———— purpureum.	Purple Eupatorium
———— maculatum.	Spotted ditto
———— perfoliatum.	Perfoliate ditto
———— cæleftinum.	Blue-flowered ditto
———— aromaticum.	Aromatic ditto
———— ageratoides.	Nettle leaved ditto
Chryfocoma Lynofyris.	German goldy-locks
———— biflora.	Two-flowered ditto
Athanafia maritima.	Sea Athanafia

SYNGENESIA POLYGAMIA SUPERFLUA.

Tanacetum vulgare.	Common tanfy
———————— *v. fol. crifp.*	Curled ditto
———— myriophyllum.	
———— Balfamita.	Coft-mary

R   Artemifia

Artemisia Abrotanum.	Common southern-wood
———— santonica.	Tartarian ditto
———— campestris.	Field ditto
———— maritima.	Sea wormwood
———— glacialis.	Silky ditto
———— spicata.	Spiked ditto
———— Pontica.	Roman ditto
———— annua.	Annual ditto
———— Absinthium.	Common ditto
———— vulgaris.	Mug-wort
——————— v. fol. varieg.	Variegated ditto
———— rupestris.	Creeping ditto
———— filiginoides.	Downy southernwood
———— cærulescens.	Lavender-leaved worm-wood
———— Dracunculus.	Tarragon
Gnaphalium luteo-album.	Jersey cud-weed
———— fœtidum.	Strong scented everlasting
———— margaritaceum.	American ditto
———— dioicum v. mas.	Mountain ditto
——————— v. fœm.	Female mountain ditto
———— plantagineum.	Plantain-leaved ditto
———— erectum.	Upright ditto
———— uliginosum.	Marsh ditto
———— glomeratum.	
Xeranthemum annuum	
v. fl. purp.	Purple Xeranthemum
——————— v. fl. alb.	White ditto
Conyza squarrosa.	Great flea-bane
———— linifolia.	Flax-leaved ditto
———— asteroides.	Star-wort ditto
———— bifrons.	Oval-leaved ditto
Erigeron Canadensis.	Canadian Erigeron
———— Philadelphicus.	Spreading ditto
———— purpureus.	Purple ditto
———— acer.	Blue ditto
———— alpinus.	Alpine ditto

Erigeron

Erigeron serpentarium.	Creeping ditto
Tussilago alpina	Alpine colt's-foot
—— sylvestris.	
—— palmata.	Palmated ditto
—— Farfara.	Common ditto
—— paradoxa.	Downy-leaved ditto
—— frigida.	
—— alba.	White ditto
—— hybrida.	Long-stalked ditto, or butter-bur
—— Petasites.	Great colt's foot, or butter-bur
Senecio vulgaris.	Groundsel
—— Arabicus.	Arabian ditto
—— triflorus.	Three-flowered ditto
—— lividus.	Dingy ditto
—— viscosus.	Stinking ditto
—— sylvaticus.	Mountain ditto
—— elegans *v. fl. simp. rub.*	Single purple Jacobæa
—— —— *v. fl. plen.*	Double ditto
—— —— *v. fl. alb. simp.*	Single white ditto
—— —— *v. fl. plen.*	Double white ditto
—— erucifolius.	Hoary perennial groundsel
—— abrotanifolius.	Southern-wood leaved ditto
—— Jacobæa.	Rag-wort ditto
—— —— *v. tomentos.*	Woolly ditto
—— palustris.	Marsh ditto
—— paludosus.	Bird's tongue
—— —— *v. coriaceus.*	Thick-leaved groundsel
—— Saracenicus.	Creeping ditto
—— Doria.	Broad-leaved ditto
—— Doronicum.	Alpine ditto
Boltonia asteroides.	Starwort-flowered Boltonia
Aster alpinus.	Alpine star wort
—— Sibiricus.	Siberian ditto
—— Tripolium.	Sea ditto
—— amellus.	Plain-leaved Italian ditto

After hyſſopifolius.	Hyſſop-leaved ſtar wort
—— dumoſus.	Buſhy ditto
—— ericoides.	Heath-leaved ditto
—— linarifolius.	Savory-leaved ditto
—— linifolius.	Flax-leaved ditto
—— concolor.	Woolly ditto
—— rigidus.	Rigid ditto
—— Novæ-anglica.	New England cluſtered ditto
—— undulatus.	Wave-leaved ditto
—— grandiflorus.	Great flowered ditto
—— paludoſus.	Marſh ditto
—— patens.	Spreading hairy ſtalked ditto
—— folioloſus.	Leafy ditto
—— multiflorus.	Late flowering ſmall leaved ditto
—— ſalicifolius.	Willow-leaved ditto
—— œſtivus.	Labrador ditto
—— Tradeſcanti *v. fl. cærul.*	Tradeſcants dwarf ditto
———— *v. fl. alb.*	Tradeſcants tall ditto
—— pendulus.	Pendulous ditto
—— diffuſus.	Diffuſe ditto
—— divergens.	Spreading downy ſtalked ditto
—— mutabilis.	Variable ditto
—— Novi-belgii.	New Holland ditto.
—— lævis.	Smooth ditto
—— paniculatus.	Smooth ſtalked ditto
—— corymboſus.	Heart-leaved purple ditto
—— cordifolius.	Heart-leaved ditto
—— mac{k}rophyllus.	Broad-leaved ditto
—— Chinenſis.	Chineſe ditto
—— Puniceus *v. caule purp.*	Tall purple ſtalked ditto
———— *v. tripedalis.*	Dwarf purple ſtalked ditto
—— ſpeɛtabilis.	Shewy ditto
—— tardiflorus.	Spear-leaved ditto
—— umbellatus.	Umbeled ditto
—— annuus.	American annual ditto
—— Radula.	Rough ditto

After

After mifer. — Small white ditto
—— elegans. — Elegant ditto
Solidago Canadenfis. — Common Canadian golden rod
—— fempervirens. — Narrow-leaved evergreen ditto
—— reflexa. — Reflexed ditto
—— lateriflora. — Lateral-flowered ditto
—— afpera. — Rough-leaved ditto
—— altiffima. — Tall ditto
—————— *v. rugof.* — Wrinkled-leaved ditto
—— nemoralis. — Woolly-ftalked ditto
—— arguta. — Sharp notched ditto
—— elliptica. — Oval-leaved ditto
—— odora. — Sweet fcented ditto
—— lanceolata. — Grafs-leaved ditto
—— lævigata. — Flefhy-leaved ditto
—— Mexicana. — Mexican golden-rod
—— viminea. — Twiggy ditto
—— ftricta. — Willow-leaved ditto
—— bicolor. — Two coloured ditto
—— rigida. — Hard-leaved ditto
—— cæfia. — Maryland ditto
—— flexicaulis. — Crooked-ftalked ditto
—— ambigua. — Angular-ftalked ditto
—— Virga-aurea. — Common ditto
—— Cambrica. — Welfh ditto
Cineraria cordifolia. — Heart-leaved Cineraria
—— alpina. — Mountain ditto
—— paluftris. — Marfh ditto
—— maritima. — Sea rag-wort
Inula Helenium. — Elecampane
—— dyfenterica. — Middle flea-bane
—— pulicaria. — Small Inula
—— fquarrofa. — Net-leaved ditto
—— falicina. — Willow-leaved ditto
—— crithmoides. — Golden famphire
—— Æftuans. — Oval-leaved ditto

Inula hirta.	Hairy Inula
Arnica montana.	Mountain Arnica
——— fcorpioides.	Alternate-leaved ditto
Doronicum Pardalianches.	Great leopard's-bane
————. plantagineum.	Plaintain-leaved ditto
——— Bellidiaſtrum.	Daiſy-leaved ditto
Helenium autumnale.	Smooth Helenium
Bellis perennis *v. ſylv.*	Common daiſy
——— —— *v. ſl. plen. rub.*	Double red ditto
——— —— *v. ſl. plen alb.*	Double white ditto
——— —— *v. ſl. plen. varieg.*	Double flowered variegated do.
——— —— *v. ſl. plen. fiſtuloſ.*	Double quilled ditto
——— —— *v. prolifer.*	Hen and chicken ditto
——— Luſitanica.	Portugal ditto
Bellium minutum.	Baſtard daiſy
Tagetes patula *v. ſl. ſimp.*	Single flowered French mary-gold
——— —— *v. ſl. plen.*	Double ditto
——— —— *v. ſl. varieg. ſimp.*	Single flowered ſtriped ditto
——— —— *v. ſl. varieg. plen.*	Double ditto
——— erecta *v. ſl. ſimp.*	Single flowered African ditto
——— —— *v. ſl. pleno.*	Double ditto
——— —— *v. fiſtuloſ.*	Quilled ditto
Zinnia pauciflora.	Yellow Zinnia
——— multiflora.	Red ditto
——— verticillata.	
Chryſanthemum ſerotinum.	Creeping rooted Chryſanthe-mum
——— —— alpinum.	Alpine ditto
——— —— Leucanthemum.	Oxe-eye daiſy
——— —— montanum.	Mountain ditto
——— —— inodorum.	Scentleſs ditto
——— —— corymboſum.	Mountain Chryſanthemum
——— —— Indicum.	Indian ditto
——— —— millefoliatum.	Milfoil leaved ditto
	Chryſanthemum

Chryfanthemum coronarium
     v. fimp. Common garden ditto
———— ————→ *v. fl. plen.* Double flowered ditto
———— ———— *v. fiftulof.* Quilled ditto
———————fegetum. Corn marygold
———— ————tricolor. Three coloured ditto
Madia mellita.
Matricaria Parthenium *v. fl.*
       fimp. Common feverfew
———————— *v. fl. plen.* Double ditto
———— maritima. Sea ditto
———— Chamomilla. Corn ditto
———— fauveolens. Sweet ditto
Anthemis maritima. Sea camomile
———— Chia. Cut-leaved ditto
———— nobilis *v. fl. fimp.* Garden ditto
———— *v. fl. plen.* Double-flowered ditto
———— arvenfis. Corn ditto
———— Auftriaca. Auftrian ditto
———— Cotula. Stinking ditto
———— Pyrethrum. Spanifh ditto
———— tinctoria. Yellow ditto
Achillea Santolina. Lavender cotton-leaved milfoil
———— Ageratum. Sweet ditto
———— falcata.
———— tomentofa. Woolly ditto
———— pubefcens. Downy ditto
———— abrotanifolia. Southern wood leaved ditto
———— macrophylla. Feverfew-leaved ditto
———— impatiens.
———— Clavennæ. Silvery-leaved ditto
———— Ptarmica *v. fl. fimp.* Common fneeze-wort
———————— *v. fl. plen.* Double flowered ditto
———— ferrata. Notch-leaved milfoil
———— alpina. Alpine ditto
———— atrata. Black cuped ditto
         Achillea

Achillea magna.	Great ditto
———— Græca.	
———— Millefolium *v. fl. alb.*	Common yarrow
——————— *v. fl. purp.*	Purple ditto
———— nobilis.	Shewy ditto
Sigefbeckia Orientalis.	Oriental Sigefbeckia
————Occidentalis.	American ditto
Bupthalmum maritimum.	Sea ox-eye
——— —— falicifolium.	Willow-leaved ditto
——— —— helianthoides.	Sun flower leaved ditto

SYNGENESIA POLYGAMIA FRUSTRANEA.

Helianthus annuus *v. fl. fimp.*	Common annual fun-flower
——————— *v fl. plen.*	Double flowered ditto
———— multiflorus *v. fimp.*	Many flowered perennial ditto
——— — *v fl. plen.*	Double flowered ditto
———— tuberofus.	Jerufalem artichoke
———— decapetalus.	Ten petaled fun-flower
———— giganteus.	Gigantic ditto
———— altiffimus.	Tall ditto
———— divaricatus.	Rough-leaved ditto
Rudbeckia laciniata.	Broad Jagged leaved Rudbeckia
———— digitata.	Narrow ditto
———— hirta.	Hairy ditto
———— fulgida.	Bright ditto
———— purpurea.	Purple ditto
———— amplexicaulis.	Stem clafping ditto
Ximinefia enfeloides.	Hoary leaved Ximinefia
Coreopfis verticillata.	Whorled-leaved Coreopfis
———— minor.	Small ditto
———— tripteris.	Three-leaved ditto
———— aurea.	Hemp-leaved ditto
———— auriculata.	Ear-leaved ditto
———— alternifolia.	Alternate-leaved ditto

Coreopfis

Coreopfis procera.	Tall ditto
Centaurea Crupina.	Black feeded centaury
—— mofchata.	Purple fweet fultan
—— v. Amberboi.	Yellow ditto
—— v. fl. alb.	White ditto
—— alpina.	Alpine centaury
—— Centaurium.	Great ditto
—— Phrygia.	Auftrian ditto
—— nigra.	Black ditto
—— pullata.	Various coloured ditto
—— montana v. latifol.	Broad-leaved mountain ditto
—— v. anguft.	Narrow-leaved ditto
—— Cyanus.	Corn ditto
—— paniculata.	Panicled ditto
—— Sibirica.	Siberian ditto
—— fempervirens.	Evergreen ditto
—— Scabiofa.	Greater knapweed
—— Orientalis.	Oriental centaury
—— Jacea.	Common ditto
—— Rhapontica.	Swifs ditto
—— glaftifolia.	Woad-leaved ditto
—— fonchifolia.	Sow-thiftle leaved ditto
—— Ifnardi.	Jerfey ftar thiftle
—— benedicta.	Bleffed ditto
—— eriophora.	Woolly-headed centaury
—— Calcitrapa.	Common ftar ditto
—— folftitialis.	St. Barnaby's thiftle
—— melitenfis.	Clufter-headed Centaurea
—— ficula.	Brown fpined ditto
—— Verutum.	Dwarf ditto
—— Salmantica.	Lyre-leaved ditto
—— muricata.	Purple flowered ditto
—— aurea.	Great golden ditto

SYNGENESIA POLYGAMIA NECESSARIA.

Silphium laciniatum.	Jagged-leaved Silphium
—— terebinthum.	Broad-leaved ditto
—— perfoliatum.	Square-ftalked ditto
—— vaginatum.	
—— trifoliatum.	Three-leaved ditto
Alcina perfoliata.	Perfoliate Alcine
Polymnia Uvedalia.	Broad-leaved Polymnia.
Calendula arvenfis.	Field mary gold
—— officinalis *v. fl. fimp.*	Common ditto
—————— *v. fl. alb.*	Double flowered ditto
—— pluvialis.	Small cape ditto
—— hybrida.	Great cape ditto
Arctotis anthemoides.	
Filago Germanica.	Common cudweed
—— montana.	Leaft ditto
—— Gallica.	Corn ditto
—— Leontopodium.	Lions-foot

SYNGENESIA POLYGAMIA SEGREGATA.

Echinops fphærocephalus.	Great globe thiftle
—— Ritro.	Small ditto

SYNGENESIA MONOGAMIA.

Jafione montana.	Annual fheeps fcabious
——————— *v. perennis.*	Perennial ditto
Lobelia Dortmanna.	Water Lobelia
—— Cardinalis.	Scarlet ditto, or cardinal flower
—— fiphilitica.	Blue ditto
—— inftata.	Bladder podded ditto
—— Cliffortiana.	Purple ditto
——————— *v. urens.*	Stringy ditto
—— minuta.	Leaft ditto

Viola

Viola palmata.	Palmated violet
—— pedata.	Multifid ditto
—— fagittata.	Arrow-leaved ditto
—— lanceolata.	Spear-leaved ditto
—— obliqua.	Oblique-flowered ditto
—— cucullata.	Hollow-leaved ditto
—— hirta.	Hairy ditto
—— paluftris.	Marfh ditto
—— odorata *v.fl. purp.*	Single purple fweet ditto
——— *v.fl. plen.*	Double purple ditto
——— *v.fl. alb. fimp.*	Single white ditto
——— *v.fl. plen.*	Double white ditto
—— pilofa.	
—— canina *v.fl. cærul.*	Blue flowered dog's violet
——— *v. alb.*	White flowered ditto
—— montana.	Mountain violet
—— ftriata.	Striated ditto
—— pubefcens.	Downy ditto
—— mirabilis.	Broad-leaved ditto
—— biflora.	Two flowered ditto
—— uniflora.	Siberian ditto
—— lutea.	Yellow ditto
—— tricolor.	Hearts-eafe
——— *v. bicolor.*	Panfies
——— *v. lutea.*	Yellow-flowered hearts-eafe
—— grandiflora.	Great yellow violet
—— Zoyfii.	
—— calcarata.	Alpine ditto
—— cornuta.	Pyrenean ditto
Impatiens Balfimina.	Garden balfam
—— Noli-tangere.	Touch-me-not

CLASSIS XX.

GYNANDRIA DIANDRIA.

Orchis bifolius.	Butterfly Orchis
——— pyramidalis.	Pyramidal ditto
——— Morio.	Female ditto
——— mafculus.	Male ditto
——— latifolius.	Broad-leaved ditto
——— maculatus.	Spotted ditto
——— conopfeus.	Red handed ditto
Satyrium viride.	Frog fatyrion
——— albidum.	White ditto
Ophrys ovata.	Tway blade
——— fpiralis.	Lady's traces
——— Monorchis.	Mufk Ophrys
——— mucifera.	Fly ditto
——— apifera.	Bee ditto
Serapias longifolia.	Long-leaved helloborine
——— grandiflora.	Great-flowered ditto

GYNANDRIA TRIANDRIA.

Sifyrinchium Bermudianum.	Small Sifyrinchium
——— ftriatum.	

GYNANDRIA HEXANDRIA.

Ariftolochia rotunda.	Round rooted birthwort
——— Clematitis.	Upright ditto

GYNANDRIA POLYANDRIA.

Arum Dracunculus.	Common dragon
——— Dracontinum.	Green ditto
——— triphyllum.	Three-leaved green-ftalked ditto

Arum

Arum maculatum *v. fol. macul.*	Common spotted Arum
———— *v. fol. non macul.*	Unspotted ditto
—— Italicum.	Italian ditto
—— tenuifolium.	Grass-leaved ditto
Calla palustris.	Marsh Calla
Zostera marina.	Sea grass-weed

CLASSIS XXI.

MONOECIA MONANDRIA.

Zannichellia palustris.	Marsh Zannichellia
Chara tomentosa.	Brittle stonewort
—— vulgaris.	Stinking ditto
—— hispida.	Prickly ditto
—— flaxilis.	Smooth ditto

MONOECIA DIANDRIA.

Lemna trisulca.	Ivy-leaved duck's meat
——— minor.	Least ditto
——— gibba;	Gibbous ditto
——— polyrrhiza.	Greater ditto

MONOECIA TRIANDRIA.

Typha latifolia.	Great cat's-tail
——— angustifolia.	Narrow-leaved ditto
Sparganium erectum.	Great bur-reed
——— simplex.	Small ditto
Tripsacum dactyloides.	Rough-seeded Tripsacum
Carex dioica.	Small Carex
—— pulicaris.	Flea ditto
——— stellulata.	Starry ditto
—— curta.	White ditto
—— ovalis.	Naked ditto
——— remota.	Remote ditto

Carex

Carex axillaris.

———— incurva.	Curved Carex
———— arenaria.	Sea ditto
———— intermædia.	Soft ditto
———— divifa.	Marfh ditto
———— muricata.	Prickly ditto
———— divulfa.	Gray ditto
———— vulpinā.	Great ditto
———— paniculata.	Panicled ditto
———— tereitufcula.	Round ftalked ditto
———— digitata.	Digitated ditto
———— clandeftina.	
———— pendula.	Pendulous ditto
———— ftrigofa.	Loofe ditto
———— præcox.	Vernal ditto
———— filiformis.	Downy ditto
———— flava.	Yellow ditto
———— fulva.	
———— extenfa.	Extended ditto
———— diftans.	Diftant flowering ditto
———— panicea.	Pink ditto
———— capillaris.	Capillary ditto
———— depauperata.	Charleton ditto
———— fylvatica.	Wood ditto
———— recurva.	Heath ditto
———— pallefcens.	Pale ditto
———— limofa.	Brown ditto
———— Pfuedo-Cyperus.	Baftard ditto
———— atrata.	Black ditto
———— pilulifera.	Ball-bearing ditto
———— rigida.	Rigid ditto
———— cæfpitofa.	Turfy ditto
———— ftricta.	Stiff ditto
———— riparia.	Common ditto
———— paludofa.	Acute ditto
———— acuta.	Slender fpiked ditto

Carex

Carex veficaria.	Bladder Carex
———— ampullacea.	Beaked ditto
———— hirta.	Hairy ditto

MONOECIA TETRANDRIA.

Littorella lacuftris.	Small Littorella
Urtica pilulifera.	Roman nettle
———— Dodartii.	Pellitory-leaved ditto
———— urens.	Leffer ditto
———— dioica.	Common ditto
———— Canadenfis.	Canada ditto

MONOECIA PENTANDRIA.

Xanthium Strumarium.	Leffer burdock
Parthenium integrifolium.	Entire-leaved Parthenium
Amaranthus tricolor.	Three coloured Amaranthus
———— lividus.	Livid ditto
———— Blitum.	Leaft ditto
———— viridis.	Green ditto
———— oleraceus.	Eatable ditto
———— hybridus.	Cluftered ditto
———— flavus.	Pale ditto
———— retroflexus.	Hairy ditto
———— hypocondriacus.	Prince's feather
———— caudatus.	Love-lies-bleeding

MONOECIA POLYANDRIA.

Ceratophyllum demerfum.	Prickly-feeded hornwort
Myriophyllum fpicatum.	Spiked water milfoil
———— verticillatum.	Whorled ditto
Sagittaria fagittifolia.	Common arrow-head
Poterium Sanguiforba.	Common burnet
———— hybridum.	Sweet ditto

MONOECIA

MONOECIA SYNGENESIA.

Momordica Elaterium.	Squirting cucumber
Cucurbita Pepo	Pumpkin-gourd
——— verrucofa.	Warted ditto
——— Melopepo.	Squafh ditto
Cucumis Melo.	Common melon
——— fativus,	Common cucumber
Bryonia dioica.	Red berried briony

CLASSIS XXII.

DIOECIA PENTANDRIA.

Spinacia oleracea *v. femin-læv.*	Smooth fpinage
——— *v. femin-fpinof.*	Prickly ditto
Cannabis fativa.	Common hemp
Humulus Lupulus.	Hops

DIOECIA HEXANDRIA.

Tamus communis.	Black briony
Rhodiola Rofea.	Rofe root
Diofcorea villofa.	Hairy Diofcorea

DIOECIA ENNEANDRIA.

Mercurialis perennis.	Dog's mercury
——— annua.	Annual ditto
Hydrocharis Morfus-ranæ.	Frog-bit

DIOECIA DODECANDRIA.

Datifca cannabina.	Baftard hemp

DIOECIA MONADELPHIA.

Napæa lævis.	Smooth Napæa
—— fcabra.	Rough ditto

CLASSIS

CLASSIS XXIII.

POLYGAMIA MONOECIA.

Veratrum album.	White hellebore
———— nigrum.	Dark-flowered Veratrum
———— luteum.	Yellow-flowered ditto
———— viride.	Green-flowered ditto
Holcus mollis,	Creeping foft grafs
——— lanatus.	Meadow ditto
——— odoratus.	Sweet fcented Holcus
Ægilops ovata.	Oval fpiked hard-grafs
——— triuncialis.	Long fpiked ditto
Valentia cruciata.	Common crofs-wort
——— glabra.	Smooth ditto
Parietaria officinalis.	Wall pellitory
Atriplex portulacoides.	Common fea purflane
———— hortenfis *v. virid.*	Green garden orache
———————— *v. rubr.*	Red ditto
——— laciniata.	Jagged fea ditto
——— haftata.	Wild ditto, or lamb's-quarters
——— patula.	Spreading ditto
——— littoralis.	Grafs-leaved ditto
——— ferrata.	Serrated ditto

P. S. Eighty-fix parcels of feeds of plants from the Eaft Indies, have been prefented to the Society, by a member, fince the foregoing catalogue was printed. A particular account of thefe fhall be given in our next volume.

Printed in the United States
By Bookmasters